Iron Has Memory, Rocks Breathe Slowly, Crystals Learn

Long Term Thinking and Cultural Change—
Revisited

Iron Has Memory, Rocks Breathe Slowly, Crystals Learn

Long Term Thinking and Cultural Change—
Revisited

Keith Q. Owen,
A. Steven Dietz, &
Robert "Skip" Culbertson

EMERGENT™
P U B L I C A T I O N S

Photo Credits

GOLIAD V: Photo credit Michael P. Dietz.

Mother & Child: Limestone sculpture by Steven Dietz; Photo credit Terry M. Rains.

Iron Pyrite: Private collection of and photo credit Terry M. Rains.

Iron Has Memory, Rocks Breathe Slowly, Crystals Learn: Long Term Thinking and Cultural Change—Revisited
Written by: Keith Q. Owen, A. Steven Dietz, & Robert "Skip" Culbertson

Library of Congress Control Number: 2014932495

ISBN: 978-1-938158-12-4

Copyright © 2014
Emergent Publications,
3810 N. 188th Ave, Litchfield Park, AZ 85340, USA

Printed in the United States of America

Contents

Preface

In 1992, Oscar Mink was asked to present the keynote address at an Australian Quality Council HAMM awards in Sydney, Australia. He chose to develop a presentation that reflected his ideas about open systems and organizations, and the fairly new idea of fractals as they relate to growth and change. As a part of his presentation he and a small group of friends developed three metaphors to carry their ideas—iron, rocks and crystals. Following the presentation Oscar, Keith Owen and Steve Bright wrote a paper with the same title as this paper and published it in the Asian Pacific Journal of Quality Management in 1993. Unfortunately that journal had a very limited life and finding a copy of the original article is difficult. That being said, the ideas captured by Oscar, Keith and Steve found life through Oscar's students at the University of Texas at Austin and through their work consulting with a variety of organizations. Oscar passed away in September of 2004 and Steve Bright seems to have found a good hiding place down under. Keith, on the other hand, has continued to write and work in the field of organizational dynamics. It was Keith's idea to resurrect these metaphors and see how they reflect organizational life 20 years after their inception.

Keith, Steven & Skip—2013

Synopsis

I n the pattern of stability, there lie the seeds of chaos and change; and in the midst of chaos, there are positive forces which can create order and stability. Effective people and effective organizations come to grips with and learn to value both chaos and order as being essential to coping with complexity and growth. If we, as humans, are to thrive, we must design and develop processes that enable the individual, the work group, and the organization to embrace these contradictory yet complementary states. This paper presents a systems model that speaks to the power and productivity of order and to the energy of chaos and transformation.

Two interesting concepts will help frame our introduction to this paper. First, the preface to one of the last hard-cover editions released by Encyclopedia Britannica suggested that of all the scientists, inventors, artists, and professional thinkers in the history of the world, over ninety-nine percent of them are alive today. Second, Paul Reber reported the that the human brain memory storage capacity of something close to 2.5 petabytes or roughly three million hours of TV shows—or the equivalent of a TV running continuously for 300 years (Reber, 2010).

We are not sure whether both of these comments are valid. Perhaps they are apocryphal; perhaps they are merely extrapolations of a trend. Whether they are precisely true or not, these comments illustrate two essential facts:

1. The world is a very complex place.

2. Humans have the capacity and are very capable of understanding complexity.

Human development is about attempts to apprehend complexity and synthesize it, in order to create a coherent picture which can be communicated to others with relative simplicity and accuracy. Organizations do the same: in their marketing, products, development, accounting processes, strategic planning, manufacturing, and human resource management. Our organizations constantly strive to understand the forces surrounding them, so as to monitor, predict, and ultimately adapt to them. This monograph is our attempt to shed some new light on how to master the forces of complexity.

Part 1: Iron has memory

The myth of fingerprints,
Over the mountains, down the valley
Lives the former talk show host
And far and wide his name was known.
He said, there's no doubt about it,
It was the myth of fingerprints.
That's why we must learn to live alone.

Paul Simon. All around the World

Culture is inevitable

All fingerprints exhibit variation: they're all different. Therefore we are all different. Yet we all possess fingerprints, therefore we all the same. We are both all the same and all different. Most of us don't have a great deal of difficulty with a concept such as this.

The apparent paradox is, of course, at the heart of quality. There is always variation yet processes which produce variation are understandable and controllable. Nevertheless, all managers undertaking a transformation process will experience a massive, over-arching complexity when they attempt to comprehend and adapt the systems within which they work. We call this complexity culture. At the level of imagery, culture is both the ties that bind us and the walls that separate us. It shows us that we are all the same, while constantly reminding us that we are all different. It tells us that we are part of the group, while never failing to show us that we are alone.

This is the story of how we begin to remember. Who will survive? Companies that adapt constancy of purpose for quality, productivity, and service, and go about it with intelligence and perseverance, have a chance to survive. Of course, they must offer products and services that have a market. Charles Darwin's law of the survival of the fittest, and that the unfit do not survive, holds in free enterprise as well as in nature's selection.

- Why do some companies outperform other companies in the long and the short term?

- Is there a relationship between culture and the performance of the organization?

- What, if anything, distinguishes such high performing companies from low performing companies?

- Are there measurable features characteristic of high performing companies? And if so, can they be described and manipulated?

- What can be done to transform a culture from one that is performing below the level of excellence to one that consistently achieves high levels of excellence?

- What is the role of leadership in such high performing companies?

What we hope to accomplish in this short book is an exploration of some of the pertinent issues surrounding what culture is and how it functions to affect the performance of organizations. Namely:

- The attributes of high performing cultures,

- The observable features of such cultures,

- The role of leadership in bringing about cultural change, and

- Strategies for organizational and cultural transformation.

We couch this discussion within the framework of organizations as complex adaptive systems. Before getting into the meat of the discussion, let's develop agreement on what we mean by the term culture in an organizational context.

What is culture?

A review of the relevant literature shows that there are as many definitions of culture as there are writers. Early on, Kotter and Heskett (1992) pointed out that the term culture was coined to represent in a holistic sense the qualities of a specific human group which serve to distinguish it from other groups and are passed (relatively intact) from one generation to the next. The essence of what culture and its effects are is reflected in what a Caribou Eskimo is quoted as saying to a government official, who wanted the tribes to restrict their nomadic travels—"What can we do? We were born with the Great Unrest. Our father taught us that life is one long journey on which only the fittest survive."

A definition

What then is culture? From the organizational perspective, culture is the set of shared beliefs, assumptions, and values (conscious and unconscious) which serve to govern behavior (shared collective practices) in an organization. Most definitions suggest that culture evolves from shared beliefs about:

- What is reality;

- What is possible in that reality; and

- What is important and worth striving for?

Culture represents the shared beliefs and assumptions that determine what a group of employees think is important (values), what they believe to be possible, and what they perceive to be the reality of their situation. The Eskimo described above sounds odd, anachronistic, and out-of-touch with reality. Yet, he is a prisoner of the only reality he knows—that which he inherited from his father—his culture.

Consider how this relates to large organizations. The corporate world is littered with the corpses of organizations that were unable, unwilling, or both to understand the need for a change in the set of assumptions they used to define what was real, what was possible, and what was important. All too often the necessity for change was forced upon them by customers and competitors. Like the Eskimo, they knew only one way to respond, and as their context had changed, it turned out to be the wrong way. Their inability or unwillingness to adapt became their death knell.

The structure of magic—Or, how cultures develop

Organizations develop cultures in response to environmental conditions. Culture emerges as people get together for the purpose of meeting their shared needs or solving problems that are of mutual concern. George Homans (1961) depicted the emergence of culture as the outcome of activities engaged in by people to meet their shared needs, which in turn produce interactions that create sentiments, values, norms, traditions, and symbols. Thus, members of an organization have the need and significant opportunities for interacting to make and implement plans, complete work activities, resolve dilemmas, and solve problems. Over time, people in organizations evolve highly specialized problem solving methodologies which will include assumptions about how to balance individual with group concerns, how to control and yet empower people, how conflict is to be resolved, and so on. In fact, Schein (1990) describes culture as:

The pattern of basic assumptions invented, discovered, or developed by a given group as it learns how to cope with its problems of external adaptation and internal integration that has worked well enough to be considered valid and to be taught to new members as the correct way to perceive, think, and feel in relation to these problems and dilemmas.

Culture is shaped by values and meaning; and reality and is shaped by both of them. Bruce Chatwick (1987) describes his meeting with a religious leader at a Mosque in Timbuktu:

The marabout interrupted his prayers to ask me a few questions.
There is a people called the Mericans?' he asked
There is.
They say they have visited the moon.
They have.
They are blasphemers.

The cultural experience of the speaker tells him more than just what is real or possible. It tells him what is right or wrong.

Cultural beliefs and values are resistant to change and tend to persist even when the membership of the organization changes. Thus, beliefs about value, reality, and opportunity tend to persist over time and become self-perpetuating.

Levels of culture

At the visible level, culture refers to the patterns of observable behavior (often shaped by language) or the style of the organization that new employees are encouraged to emulate and follow. In fact, there are two levels of culture, the visible and invisible. These levels of interdependent and each level of culture influences the other. Thus, a high commitment to shared values influences the behavior of all employees in an organization. On the other hand, behavior also influences culture. For example, when employees begin to talk about and provide for quality customer service and meeting customer expectations, they actually change their values regarding the importance of customers (Aronson, 2008). This change in values can lead to further change in the behavior of employees toward customers, especially when this change is reinforced by increasing opportunities to talk to customers and incentives for more effectively meeting customer needs. This mutual interdependence of culture and behavior is illustrated in Exhibit 1.

How the magic works

Culture works by influencing the choices people make and then selectively rewarding those choices considered to be appropriate. One of the ways that culture influences behavior is through its effect on what people perceive, think, and feel. Psychologists, linguists, and anthropologists have convincing proof that perception is shaped by belief. What one believes to be true influences what one perceives, and what one perceives influences choice behavior. As an example, think about a popular therapeutic intervention for treating

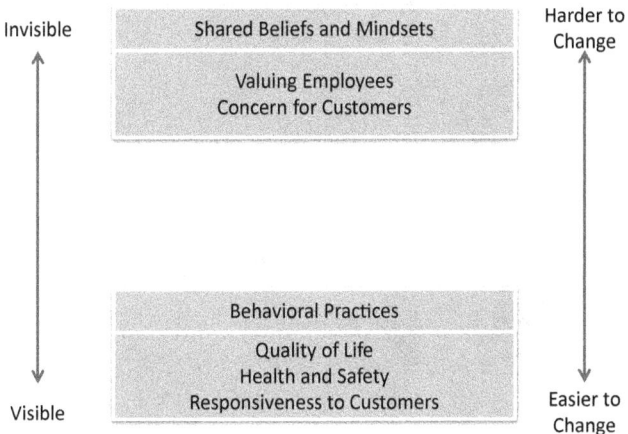

Exhibit 1 *Mutual interdependence of culture and behavior*

depression—affirmations. What you think influences how you feel, how you feel influences how you act, how you act influences what you think (Sherman, Cohen, 2006). Schein's (1990) view of the levels of culture and their effects on choice behavior is illustrated in Exhibit 2.

Exhibit 3 illustrates more clearly how culture literally creates the nature of the container that is an organization. Culture is like a mental habit and defines your being. Your way of being defines how you think and what you think about; what you think determines what you DO; and what you do creates what you have.

Routines and rewards solidify culture

The culture of an organization can be observed and understood by examining the routines and rewards that characterize it (Schneider, 1988). Routines are those things we do which are supported, nurtured, and come to be expected by the organization. Routines

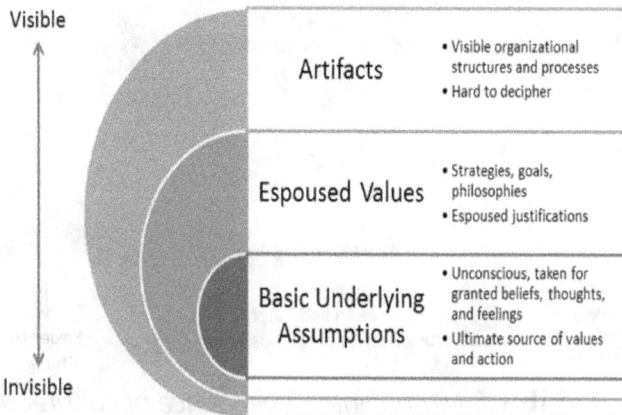

Visible		
	Artifacts	• Visible organizational structures and processes • Hard to decipher
	Espoused Values	• Strategies, goals, philosophies • Espoused justifications
Invisible	**Basic Underlying Assumptions**	• Unconscious, taken for granted beliefs, thoughts, and feelings • Ultimate source of values and action

Exhibit 2 *Levels of culture*

CULTURE & BEHAVIOUR

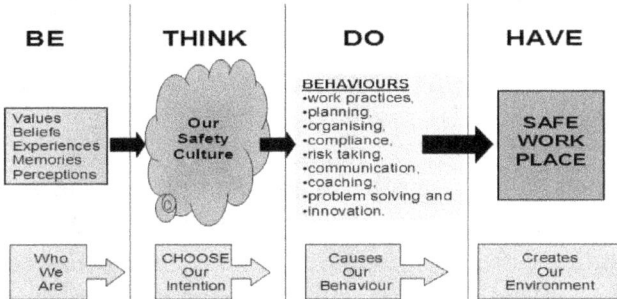

BE	THINK	DO	HAVE

BEHAVIOURS
•work practices,
•planning,
•organising,
•compliance,
•risk taking,
•communication,
•coaching,
•problem solving and
•innovation.

Values
Beliefs
Experiences
Memories
Perceptions

Our Safety Culture

SAFE WORK PLACE

| Who We Are | CHOOSE Our Intention | Causes Our Behaviour | Creates Our Environment |

Exhibit 3 *Culture shapes our behavior*

communicate to employees what they should and should not do, if they wish to remain members in good standing. And this is one important function of culture— to define boundaries for members of the culture.

We did some work with a government organization attempting to move from a traditional bureaucracy to a more customer responsive, client-based service provider. After some observation, it was noticed that middle managers hated being interrupted. If a staff member came up to them with a problem, they would often methodically complete the task they were involved in before addressing the staff member. Sometimes an employee would be standing in front of a manager's desk for several minutes. The same managers complained about the reluctance of counter staff to drop everything to serve a customer the moment he or she walked up to the service counter (Owen, Northcutt, Dietz, 2013).

Rewards are the kinds of consequences a given behavior produces in a particular setting. Behaviors which meet or exceed expectation are rewarded and recognized.

These behaviors come to serve as symbols of how to get ahead in the company. For example, 3M has, over the years, recognized and rewarded innovation, thus innovation is a highly regarded value at 3M (3M Company, 2002).

Often members are educated about the routines and rewards of the culture through stories and myths that are repeated over and over again by members. For example, at one company for whom we consult, there is a myth about people who have been fired because they took a risk and lost. In reality, only one person we knew of was actually terminated for risk taking. This was a rare case in which the risk taken was a long shot and consequences of failure so severe. Nevertheless, in this culture, individual risk taking and openness are suppressed by the oft repeated story of the mass of risk takers who risked, lost, and were callously sacked when it didn't work out as planned. This apprehension about risk taking is a fundamental problem for this organization, which is undertaking a transformation toward a culture of innovation—a transformation for which intelligent risk taking (and thus occasional failure) is a core behavior.

A large manufacturing and service provider we worked with had a problem with customer orders. Every month, from around the country, regions would supply figures that showed they were right on target for new services and equipment orders. Always! Common sense, however, tells us that there should be some variation —good months and bad months. Either that or it was an incredibly stable system. Well, it was stable, largely through the intervention of staff.

It was well known that Regional Managers were judged on the basis of monthly figures their regions returned. So what would happen was that an enterprising clerk, operating independently, but as a result of established routine and reward systems, would notice that by the end of the month, orders might be down. The clerk would use her or his computer to create several hundred imaginary orders: just enough to get the figures right. Then on the first day of the month, the same clerk would remove the bogus orders from the computer system.

The main point here is that this was NOT a planned conspiracy. There was no collusion between managers and clerks, supervisors or ordering staff. All that existed was a clearly understood set of cultural routines and rewards, reaching from the clerk to her or his supervisor, to the manager, to the Regional Manager, and right up to the Senior Executives who viewed the figures each month. Everyone wanted the figures to be right, and so they were. The most valued thing in that culture was getting the figures right.

Characteristics of culture—The myth of iron

We normally perceive iron as inert, solid, resistant to change, brutish, and hard. But, as we said at the outset, our understanding of organizational cultures and how to shape them must be in a context which allows for contradictory states to exist side by side. Iron has memory. If we take an iron bar, apply some force, and bend it, it is impossible to return it to its original shape. If the iron is cold you cannot bend it back to the original shape and the bar will bend at some point different than the original bend.

The molecular structure of iron is notionally similar to sedimentary rock: a large number of plates resting on top of each other. When we bend the iron, some of the plates break and end up at an angle to each other. When we try to straighten the bar, the edges of the plates butt up against each other, and the bar resists the attempt to bend it with a strength we did not know it possessed. If the iron is heated it can be reshaped to look like the original bar. However, when you heat the iron you reshape the plates, despite the similar outward appearance of the newly straightened bar, internally the bar is completely different.

This may be a useful analogy for those managers exploring cultural change. If you are going to apply force to change culture, make absolutely sure that you get it right the first time, because it is impossible to go back to the original culture, and it becomes increasingly more difficult to bend culture the next time.

Culture reflects basic assumptions about survival and integration

A given culture reflects two sets of needs, one about how to survive in a constantly changing environment and two about how to manage relationships and coordinate the various activities needed to ensure survival. In a sense, culture is the silent language that shapes behavior. This silent language has a number of characteristics (see Hampden-Turner, 1990, 2009), among which are the following:

Individuals make up culture

Culture originates in the individual members, usually those in roles of power and authority. Each of these people come equipped with their own assumptions about value, possibility, and reality and it is these assumptions that influence the culture that evolves.

Culture is the source of shared meanings

People who share a culture generally attach the same meaning to the events and experiences they share. In fact, later on we will argue that creating shared meanings is a fundamental building block of culture. Shared meanings provide a basis for communicating to one another, organizing experiences, taking actions, producing results, and explaining the world, whether it is positive or negative. Moreover, culture provides a means for helping people understand how to be and do together.

Culture provides a framework for making choices

Because culture provides a shared understanding of what is important and what is not important, it provides a framework for decision making that can be a powerful integrating force in an organization.

Culture affirms and rewards excellence

The values characteristic of a culture embody the hopes and dreams of a people. As they accomplish these hopes and dreams, culture affirms and gives substance to them. Also, the beliefs which define a culture can be important incentives for high performance. These values can be the source of creative tension that pulls a group toward excellence.

Cultures tend to become self-fulfilling

Once cultures take shape, the assumptions and beliefs that come to characterize it take the form of shared expectations. Since behavior follows expectations, the beliefs and assumptions that form the culture tend to become self-perpetuating. For example, if managers assume that people are not to be trusted, they treat them as such. People naturally react to this expectation behaving in ways that are consistent with the assumption. Another way to think about the self-fulfilling nature of cultures is to recognize them as cybernetic systems (Gardner & Ashby, 1970). A cybernetic system steers itself and remains on course despite obstacles and problems through feedback loops to specific aspects of the environment. Cybernetic systems produce on-going feedback that allows them to compare where they are relative to where they are going. Cultures act like a ship's compass, which indicates the direction a ship is heading. This heading is determined by feedback from wind, water currents and so on, continuing to move a ship in its intended direction. This phenomenon, like all aspects of culture, can be both bane and blessing. Well established routines can promote or inhibit organizational transformation and change.

Cultures provide members a sense of identity and continuity

People identify with the organizational culture to which they belong. This means the reputation, values and image of the organization come to represent who a given person is. The process of identification with the organization for which we work fulfills an important need—to belong to something bigger than oneself.

In identifying with the organization, people take on aspects of that culture which become an important part of how one sees and experiences oneself. In other words, cultural membership provides people with both a sense of belonging and inclusion and identity.

Cultures balance reciprocal values and needs

Organizations must constantly seek a balance between issues of control verses empowerment, quality verses cost of producing quality, stability verses change, and between leadership and management. Culture is a reflection of how these dilemmas are resolved.

Summary

People form organizations in order to accomplish a purpose. Culture is the glue that organizes their thinking, feelings, and actions and serves as a guide to how they strive to fulfill that purpose. However, once formed, cultures tend to be self-perpetuating, and in important ways, each of the characteristics of culture contributes to the success or failure of the organization. In the next section, we will explore how culture, once formed, is self-perpetuating and determines the performance of the organization.

Part 2—Rocks breathe slowly

Marco Polo describes a bridge, stone by stone.
But which is the stone that supports the bridge? Kublai
Kahn asks.
The bridge is not supported by one stone or another,
Marco answers, but by the line of the arch they form.
Kublai Kahn remains silent, reflecting.
Then he adds, Why do you speak of the stones? It is
only the arch that matters to me.
Polo answers, Without stones, there is no arch.

Italo Calvino, Invisible Cities

The metaphor of the rock slowly breathing reflects a critical function of culture—to provide the stability, meaning and predictability in the present required to keep on keeping on. However, consensus and commonality are a double edged sword. In the beginning, they provide the framework that makes sense of the shared struggle to survive; after a while, this stability or sameness can serve to impede or derail the need to learn and adapt. In Part 2, we explore the evolution of culture and the resulting variation in the kinds of culture that evolves. We then discuss the effects of cultural type on performance.

Stone is not solid and dense. It is made up of molecules and atoms in the same way we are. It breathes slowly and deeply with the same force and regularity that we know. As rocks heat in the sun, they expand: the space between grains is slightly increased, taking in air. In the cool of night, they contract, and the grains are compressed. Sometimes cracks appear: this is known as exfoliation, a cross-over term which is now used by cosmetic companies to describe the action of cleansers on human skin.

Culture is Man's response to the need for survival

Cultural development proceeds according to the laws of selection. What works gets repeated and eventually institutionalized in the mores of a group. They are the folkways of central importance, accepted without question and embody the fundamental moral views of a group. Culture embodies the folkways of a group, and the ways of living, thinking, and acting in a group. Culture is built up without conscious design, but serves as a compelling guide of conduct. These folkways contribute to the survival of the group. Once formed, a culture shapes perception, thought, emotion, and action. It acts as a self-fulfilling prophecy providing members the predictability and stability that enable its members to act with self-confidence and assurance. If I do X, Y will happen and this is good. Here we explore this relationship between cultural development and organizational performance.

Culture is the outcome of assumptions about reality, value, and possibility

As humans, we seek meaning in our lives. This is a natural tendency and reflects a fundamental need for purpose. There are many pathways to seeking meaning. The nature of the particular path a given organization takes is a function of the assumptions the founders make about the nature of reality, possibility, and value.

Assumptions about reality

Culture reflects an organization's basic assumptions about survival and integration. An important question cultures must answer is "How do we survive in an ever changing external environment?" There are five core assumptions about this environment that channel the development of a culture (Schein, 1990):

- **Vision, mission, and strategy**—gaining consensus and shared understanding of the mission. This involves balancing the needs of a number of constituencies (customers, employees, stockholders, communities, suppliers, etc.) while maintaining focus on one core vision. When there is lack of agreement between the constituents about what the core vision and values are, then the whole edifice can fail.

- **Goals**—developing consensus on goals as derived from the vision. Moving from vision to goals is not as straightforward as it seems. One has to define an abstract concept—vision—in logical, operational terms. This is why groups need a common language about how to 'create' their vision and about how to translate the strategic to the tactical. To achieve consensus on goals, the group needs a common language and shared assumptions about the basic logistical operations. It can move from something as abstract or general as a sense of vision to the concrete goals of designing, manufacturing, and selling an actual product or service within specified and agreed-upon cost and time constraints.

- **Means**—developing consensus on how to achieve the goals through structures, division of labor, rewards, etc. People can have ambiguous goals, but for anything to happen, they must agree on how to structure the organization, and how to design, finance, build, and sell products or services—in other words how work will be accomplished in the organization. From the particular pattern of these agreements will emerge not only the "style" of the organization but also the basic design of tasks, division of labor, reporting and accountability structure, rewards, incentives, and control and information systems. This structure of assumptions can provide the regularity needed to produce value, but can also serve to strangle creativity in response to the need for survival.

- **Measurement**—developing consensus on how to assess performance. Consensus must be achieved on what to measure, how to measure it and what to do when corrections are needed. The cultural elements that form around each of these issues often become the primary focus of what newcomers to the organization will be concerned with. Such measurements inevitably become linked to how each employee is doing his or her job. If members of the group hold widely divergent concepts of what to look for and how to evaluate results, they cannot decide when and how to take remedial action. For example, what is the measure of success of an organization? Short term financial gain? Long term stability? Or something else?

*   **Correction**—developing consensus on how to cope with failure. Effective remedial action requires consensus on how to gather external information, how to get that information to the right agents of the organization who can act on it, and how to alter the internal production processes to take the new information into account. Organizations can become ineffective if there is lack of consensus on any part of this information gathering and utilization cycle.

How these assumptions are worked out strongly influences the internal integration of the organization and the groups within the organization. Conversely, how well the group(s) learns to work together strongly influences the organization's ability to respond to the external environment. Culture reflects the group's efforts to cope and learn. It can be viewed as the residue of that learning process.

Assumptions about possibility

Coping and survival require figuring out how to work together for the common good. There are six processes involved here (Schein, 2010):

- **Creating a common language and conceptual categories**. If members cannot communicate with and understand each other, a group is impossible by definition.

- **Defining group boundaries and criteria for inclusion and exclusion**. The group must be able to define membership. Who is in and who is out, and by what criteria is membership determined?

- **Distributing power and status**. Every group must work out its pecking order, its criteria and rules for how someone gets, maintains, and loses power and authority. Consensus in this area is crucial to help members manage feelings of aggression, attachment, etc.

- **Defining norms of intimacy, friendship and love.** Every group must work out its "rules of the game" for peer relationships, relationships between the sexes, and the manner in which openness and intimacy are to be handled in the context of managing the organization's tasks.

Consensus in this area is crucial to help members define trust and manage feelings of affection and love.

- **Defining and allocating rewards and punishments**. Every group must know what its heroic and sinful behaviors are and must achieve consensus on what is a reward and what is a punishment.

- **Explaining the unexplainable—ideology and religion**. Every group, like every society, faces unexplainable events that must be given meaning, this is done so that members can respond to them and avoid the anxiety of dealing with the unexplainable and uncontrollable. In other words, culture provides a framework for explaining things that are unexplainable. For example, in pre-modern cultures, thunder was seen as a sign that the gods were angry. In Mayan culture, drought was seen as a sign of sin and the only way to appease the gods for such sin was human sacrifice.

Assumptions about values and truth

As groups and organizations evolve, the assumptions they develop about external adaptation and internal integration reflect deeper assumptions about more abstract general issues. Humans need to develop consensus to have any kind of society at all. If we cannot agree on what is real, how to determine the truth or falsity of something, how to measure time, how space is allocated, what human nature is, and how people should get along with each other, society is not possible in the first place (Schein, 2010).

- **Assumptions about the nature of reality and truth**. Every culture has a set of assumptions about what is real and how to determine or discover what is real. Such assumptions tell members of a group how to determine what is relevant information, how to interpret information, how to determine when they have enough of it to decide whether or not to act, and what action to take. The most important categories of culture are the assumptions made about how reality, truth, and information are defined. Reality can exist at the organizational, group, and individual levels. The test for what is real will differ according to the level—overt tests, social consensus, or individual experience. Occupations and macro cultures differ in the degree to which they rely on moralistic traditional criteria for truth as contrasted at the other extreme with pragmatic scientific criteria. Groups develop assumptions about information that determine when they feel they have enough information to make a decision. Those assumptions reflect deeper assumptions about the ultimate source of truth. What is a fact, what is information, and what is truth—each depends not only on shared knowledge of formal language but also on context and consensus.

- **Assumptions about the nature of time**. There is probably no more important category for cultural analysis than the study of how time is conceived and used in a group or organization. Time management imposes a social order and conveys status and intention. The pacing of events, the rhythms of life, the sequence in which things are done, and the duration of events are all subject to symbolic interpretation. Misinterpretations of what things mean in a temporal context are therefore extremely likely unless group members are operating from the same sets of assumptions. The main aspects of time, including (1) past, present, near-, or far-future orientation; (2) monochronicity or polychronicity; (3) planning or developmental time; (4) time horizons; and (5) symmetry of temporal activities, help define an organization, as well as assist organizational agents begin to understand how you view and use time. Time is the key to coordination, planning, and the basic organization of daily life, yet is invisible and taken for granted. Despite time coordination being central to the workings of all social orders, it is even difficult to speak about. For example, when we are late or early, we mumble apologies and possibly provide explanations, but rarely do we ask, "When did you expect me?" or "What does it mean to you when I am late (or early)?"

- **Assumptions about the nature of space**. Most leaders are not aware of how much the assumptions they take for granted are passed on in day-to-day behavior toward employees. Assumptions about space influence many things we take for granted such as how decisions get made and by whom, how offices are arranged, how much access an employee has to their boss, the manner in which deadlines are managed, and so on. While this may seem trivial, the way leaders act out their own assumptions about time and space trains their subordinates and ultimately their entire organization to accept those assumptions.

- **Assumptions about the nature of human nature, human activity, and human relationship**. Every culture has shared assumptions about what it means to be human, and what our basic instincts are. At the organizational level, the basic assumptions about human nature are often expressed most clearly in how workers and managers are viewed. These views can become deeply ingrained in the decisions made by senior managers and over time in the very fabric of the organization itself.

Culture evolves from an influence process

Cultures evolve through an influence process. Influence is all about leadership and leaders. Therefore, we must not underscore the importance of leaders in shaping culture. We have tried to capture the essence of this influence process in the exhibit below, which itself is the outcome of a multi-year research endeavor. The influence process starts with basic assumptions about reality, possibility, and value. In other words, leaders are like all of us—they are meaning producers—and create meaning in terms of their assumptions about life.

The influence process begins with the core beliefs, mindsets, and aspirations of the senior leaders (see Exhibit 4, Box 1. Executive Mind-sets) and these are the foundation of the basic meaning producing assumptions made by senior leaders. These assumptions trigger a decision process (see Exhibit 4, Box 2. Executive Decisions) about what the organization's purpose is, who gets to participate in creating this dream, the manner in which agents are developed, the degree to which those agents are empowered to participate in decision making and resource allocation, how they will be recognized and rewarded, and how their well-being will be protected. The effects of basic assumptions on these decisions may be conscious or unconscious, but their impact is revealed in the nature of these decisions, and in the nature of the leadership practices which evolve in the organization as a consequence of these very decisions.

In fact, these decisions shape the nature of the leadership practices that evolve in the organization. It is these decisions that determine who gets to become a leader, how they are developed, how they are expected to lead, and what they are expected to achieve (see Exhibit 4, Box 3. Leadership Practices). In other words, these decisions shape the very nature of the exercise of influence for the purpose of achieving results through others. Our research regarding leadership practice and team performance has shown that these practices typically fall into several dimensions:

- Empowerment and achievement—how a given leader empowers his or her people to make tactical decisions and solve problems.

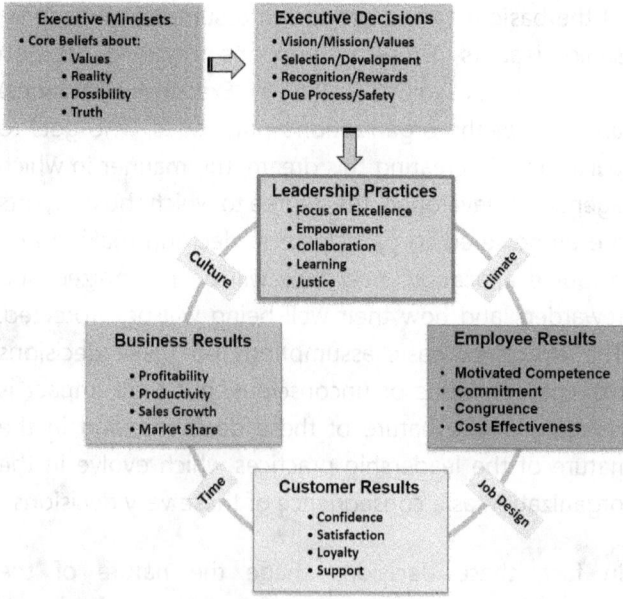

Exhibit 4 *Culture evolves from an influence process. © Somerset Consulting Group, 2014*

- Collaboration and relationship—how a given leader works with his or her people and how he or she expects his or her people to work together.

- Respect—how a given leader treats those in his charge and enables them to maintain a high level of health and safety behaviors.

- Learning—how a given leader encourages and promotes continuous learning and development of those in his or her charge.

These leadership practices produce observable and measurable effects on the level and quality of employee commitment and effort (see Exhibit 4, Box 4. Employee Response). A core outcome of leadership practice is the degree to which people are able to meet their needs and the resulting effect on employees' quality of their inner work life which evolves from success or lack thereof. Quality of inner work life refers to the flow of perceptions, feelings, thoughts, and evaluations of one's work environment. When people are enabled to meet their needs, their quality of work life is positive and they exhibit motivated effort and enthusiasm, commitment, congruence with the values of the organization, and high degrees of cost effectiveness (Amabile & Kramer, 2011). When they are not enabled to do so, their quality of inner work life becomes negative and they exhibit apathy, resignation, reduced effort and reduced performance.

Employee inner work life and correlated behaviors produce short term outcomes (see box labeled 5. Short

Term Outcomes) and these include level of customer satisfaction and loyalty, accomplishment of short term goals, and quality. Finally, these short term outcomes influence the performance of the organization (see Exhibit 4, Box 6. Long Term Outcomes). In summary, the shared meanings created by executive decisions lead to experiences that reinforce these meanings, and enable actions and long term results that are consistent with those meanings. The effects of this cycle of influence are shown in the text boxes connecting the core processes: leadership practices produce a climate that encourage and supports a given set of actions, over time creating the culture which comes to characterize the organization.

Cultural and performance

The reasoning above suggested a number of questions about cultures and performance, including:

*   Are there variables that consistently describe cultures?

*   Are variations in these variables related to the performance of the organization?

*   Can these variables be changed and if so does change reveal itself in improved performance?

To answer such questions, over a period of several years we collected data from a number of organizations using an in instrument developed by Somerset Consulting Group called CultureGRADE® (see Attachment 1)[1]. It

1. CultureGRADE® is a Registered U.S. Patent and Trademark Office

is widely accepted that employee perceptions are correlated with organization performance. Therefore, we began factor analyzing our employee survey data base and using the resulting factors to see if we could predict the organization's performance. In our initial study, census surveys of over 50,000 employees from about 50 organizations (n ranged from 400—30,000 employees) were factor analyzed and the resulting variables were then used as predictors of performance.

This analysis revealed that there were a finite set of predictor variables that accounted for about 76%—81% of the variance in our survey data. We then factored this set of variables and three broad second order constructs emerged. Exhibit 5 shows the results of this analysis.

Let's examine each of these constructs.

Shared meanings

Shared Meanings refers to the degree to which the organization is aligned around a set of core beliefs and values. Furthermore, it is the degree to which those in leadership roles reinforce these beliefs and values through their decisions and actions. The degree to which shared meanings is perceived to exist in a given organization is determined by four variables:

- Confidence in Senior Management—to what extent can senior leaders be trusted to carry out the vision of the organization.

- Identification with Vision / Mission—to what extent can I as an individual identify

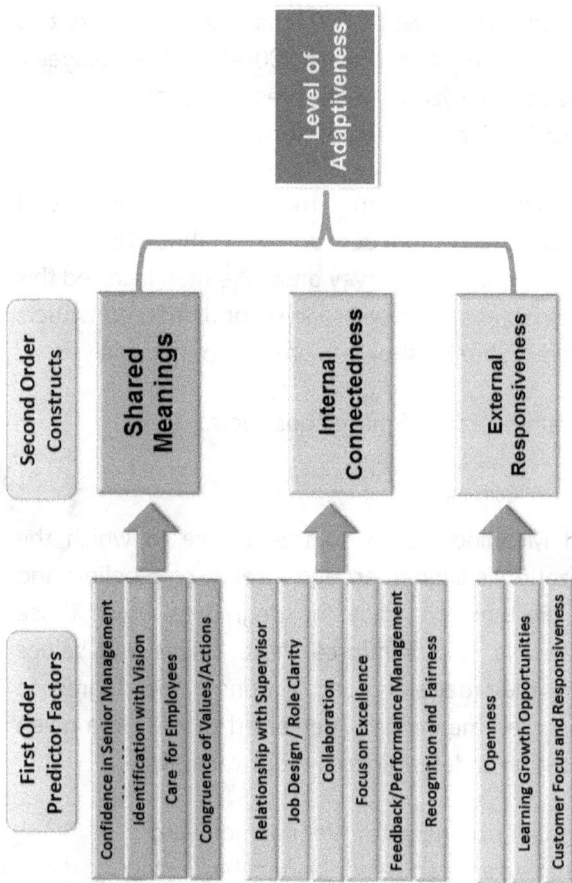

Exhibit 5 *The predictors of adaptability.* © Somerset Consulting Group, 2014.

Second Order Constructs

- Shared Meanings
- Internal Connectedness
- External Responsiveness

Level of Adaptiveness

First Order Predictor Factors

Shared Meanings:
- Confidence in Senior Management
- Identification with Vision
- Care for Employees
- Congruence of Values/Actions

Internal Connectedness:
- Relationship with Supervisor
- Job Design / Role Clarity
- Collaboration
- Focus on Excellence
- Feedback/Performance Management
- Recognition and Fairness

External Responsiveness:
- Openness
- Learning Growth Opportunities
- Customer Focus and Responsiveness

with and commit to the purposes of the organization.

- Care for Employees—to what extent does this organization care for me as a person.

- Congruence of Values and Action—to what extent do we do as we say we are going to do as an organization.

Internal connectedness

Internal connectedness refers to the degree to which employees are enabled to meet their needs through shared experiences on a daily basis. To varying degrees, shared experiences enable employees to focus on and work toward achieving the vital few priorities. The degree of internal connectedness or focus is determined by six variables:

- Relationship with Supervisor—the extent to which each employee perceives that they are cared for and trusted by their direct supervisor.

- Job Design / Role Clarity—the extent to which the jobs people are expected to perform enable them to make progress in fulfilling their needs for relationship, autonomy, mastery, and purpose.

- Collaboration—the extent to which people believe they work together as a team to achieve the shared purposes of their work group.

- Focus on Excellence—the extent to which the job provides clear standards of excellence and provides the means to achieve excellence.

- Feedback and Performance Management—the extent to which employees receive regular valid feedback about how well they are performing relative to standards.

- Recognition and Fairness—the extent to which people are recognized for their results in a fair and equitable manner.

External responsiveness

External responsiveness refers to the ability of the organization to gather and use valid information to take fit-for-purpose actions, and to provide people the opportunity to learn and grow as the context changes over time. The degree of external responsiveness is determined by three variables:

- Openness—a reflection of how able and willing the organization is to seeking out new and more effective ways of being and doing in the context in which it functions.

- Learning and Growth Opportunities— the extent to which the organization is believed to encourage and enable learning and innovation.

- Customer Focus—the extent to which the organization is focused on understanding and responding to customer needs, both internal and external.

From a conceptual point of view, scores on shared meaning, internal connectedness, and external responsiveness determine the degree to which employees feel a sense of efficacy or mastery over their environment (Bandura, 1997). We used this framework to understand the relationship between culture as defined by these predictor variables and performance. As shown by the relationship in exhibit 6, the organizations in our sample exhibited variation on the predictor variables and that this variation was strongly related to how well the organization performed. In this particular study, we tracked an index of adaptiveness (basically the score on the three broad predictor factors) of over 1,000 work units for a period of 33 months and looked at performance as a function of adaptability. As can be seen, there were substantial differences in performance among work units that were high, medium, or low on adaptability.

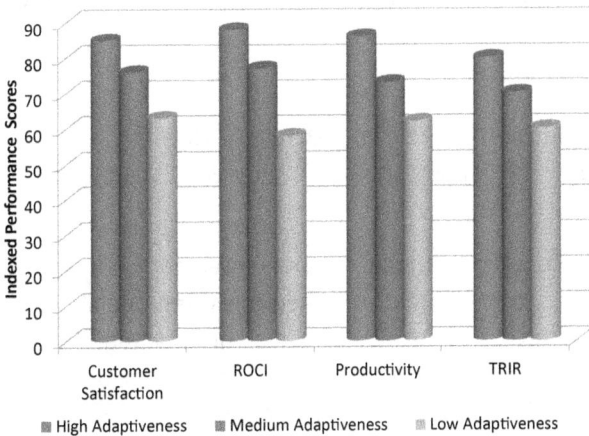

$P < 0.001$

Exhibit 6 *Adaptability and organizational performance*

A second example reflects between group differences within a single organization. Exhibit 7 shows a study looking at the financial impact of high versus low employee efficacy. The table compares certain measurements across two different groups. The first group includes all the employees working in business units that achieved top quartile employee satisfaction as compared to all business units. The second group includes all employees working in business units that achieved bottom quartile employee satisfaction as compared to all business units. These numbers demonstrate the difference in employee performance in high employee efficacy (Top 25%) and low employee efficacy business units (Bottom 25%). Top quartile employee efficacy business units earn significantly higher operating pretax profits, have lower operating expenses, and achieve higher employee retention and productivity. These are meaningful differences and show that high employee efficacy has positive financial implications.

Culture influences performance through its effects on sense of efficacy and quality of work life (QoWL)

A consistent theme in our research on the effects of culture is that it determines members' ways of perceiving, thinking, feelings, and acting. Amabile and Kramer (2011) referred to this effect as the inner work life to denote the flow of employees' perceptions, feelings, thoughts, and expectations as they experience their day to day work life. Inner work life describes the nature and quality of employees' experiences. Therefore,

	Work Climate Average	Operating Pretax %	Operating Expense as % of Sales	Workers' Comp. % of Sales	Employee Retention	Delivery Retention	Associates per 100 Thousand Cases
Top 25% work climate	4.01	7.5%	13.3	0.07	85	88	4.13
Bottom 25% work climate	3.61	5.3%	14.9	0.20	72	78	4.33

Exhibit 7 *Business units work climate impact*

it is reasonable to assume that the nature and quality of this inner life would have profound effects on the productivity of a given culture. Here we want to flesh out these effects more fully.

In terms of our influence model, leaders through their practices create experiences for employees. These experiences enable employees to meet their needs for relationship, autonomy, mastery, and purpose (Pink, 2011) to one degree or another. The psychological outcome of the relationship between these experiences and need fulfillment is a sense of efficacy (Bandura, 1997). Efficacy describes the effects of expectations on the level of achievement in a given environment. Our research has established that the scores on the CultureGRADE® provide a reliable and valid assessment of the drivers of employees' sense of efficacy, which in turn drives the outcomes in Box 4 (Exhibit 4) of our influence model, namely, commitment, effort, competence, congruence, and cost effectiveness. These drivers, working in concert with one another, form the basis of the Quality of Work Life (QoWL) Grade which is a measure of the quality of employee's inner work life.

In terms of the CultureGRADE® inner work life is very negative when the scores on the variables are uniformly low, and is very positive when the scores are uniformly high. This is important for it explains the significant correlation we have found between scores on these drivers and various measures of organizational performance including customer satisfaction, productivity, safety, quality, and timeliness.

Inner work life affects employees' level of engagement in the work of the organization. When QoWL is low, the level of engagement is determined by extrinsic factors such as fear of punishment or loss of security. When QoWL is high, level of engagement is determined by intrinsic factors such as values and aspirations.

When we start to think about why there is this strong relationship between QoWL and employees inner work life, the most plausible explanation is that QoWL is positive when employees are able to meet their needs for Relationship, Autonomy, Mastery, and Purpose at work. The way organizations enable employees to meet their needs is really quite simple: organizations provide employees a space which allows for opportunities (experiences) to meet their needs. Thus, if I give my employees the opportunities to achieve small wins on a day to day basis they develop a sense of efficacy relative to their environment. In other words, employees come to believe they are in control of their own destiny and this belief motivates them to do their work better today than yesterday. This space has three qualities which interact to encourage and motivate employees to commit to and give effort on behalf of the organization's vision and mission, namely:

1. Shared meanings,

2. Internal connectedness and focus, and

3. External responsiveness.

Cultural style is the outcome of how the cultural building blocks are put together

We like to think of these qualities as the building blocks of culture and the way in which they are put together as cultural style. When we began to look more deeply at what influences the style a culture develops. The data supported the concept that the organizations that scored high on this set of measures were those that created a work space which enabled and reinforced a high degree of efficacy. In his research on the social foundations of thought and action, Bandura (1997) discovered there are three kinds of efficacy, each of which contributes to the quality of employees' inner work life:

- Self-efficacy—this is the expectation that I have the ability to execute the tasks required of me and at the level expected of me.

- Response efficacy—this is the expectation that my efforts will lead to a result.

- Outcome efficacy (optimism)—this is the expectation or belief that my efforts make an important contribution to the success of the organization.

When people develop a sense of efficacy with regard to the organization, they develop positive feelings about and perceptions of the organization. Moreover, they will freely contribute extraordinary effort to help the organization succeed. When a sense of efficacy is lacking, they develop negative feelings about the organization and find it hard to get motivated (see Exhibit 8).

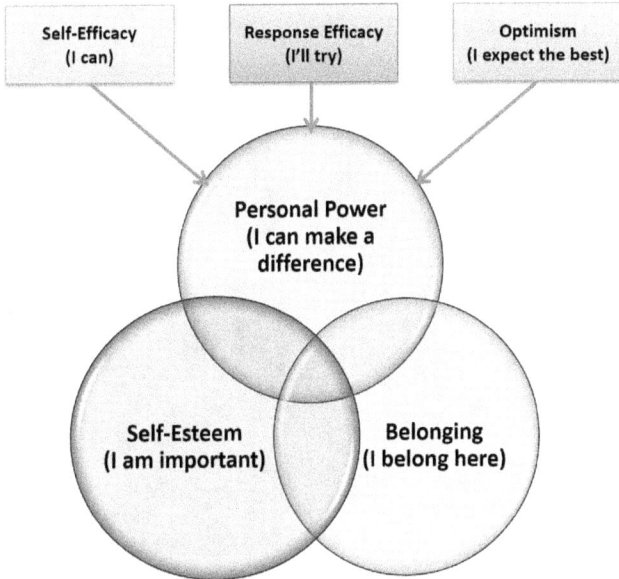

Exhibit 8 *Conditions that activate a positive quality of inner work life*

Expectations as a self-fulfilling prophecy

Efficacy is a generalized expectancy about one's ability to meet one's needs at work. This expectancy can be positive or negative. A positive expectation thus causes a positive inner work life while a negative expectation causes a negative inner work life. Quality of inner work life acts like a self-fulfilling prophecy. If members expect good things to happen in a given culture, they act accordingly; if they expect bad things to happen, they act accordingly, as illustrated in Exhibit 9.

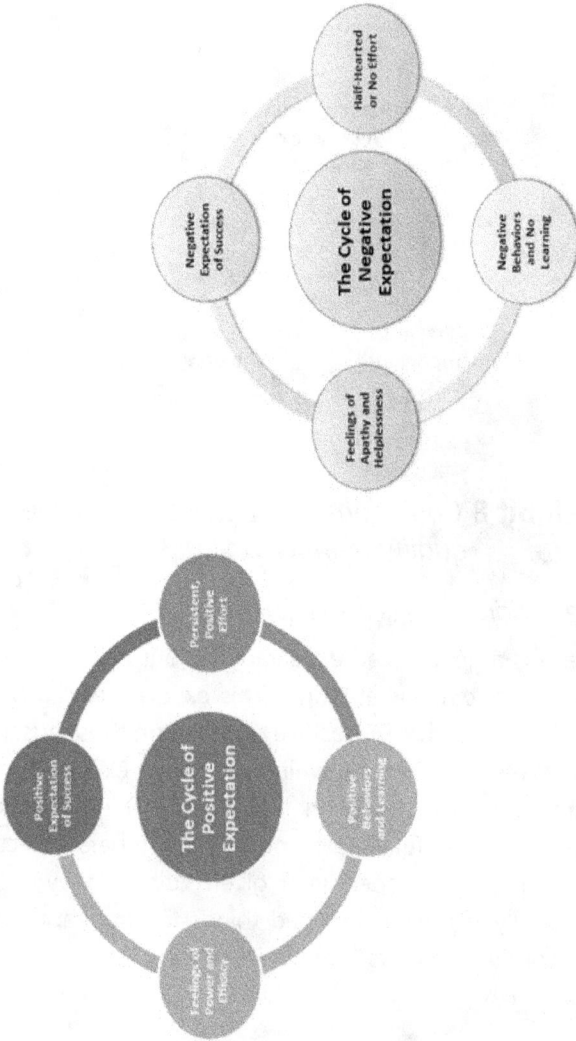

The Cycle of Positive Expectation

- Positive Expectation of Success
- Persistent, Positive Effort
- Positive Behaviors and Learning
- Feelings of Power and Efficacy

The Cycle of Negative Expectation

- Negative Expectation of Success
- Half-Hearted or No Effort
- Negative Behaviors and No Learning
- Feelings of Apathy and Helplessness

Exhibit 9 *Expectations become a self-fulfilling prophecy*

Cultural styles

Cultural style refers to the QoWL grade of a given organization at a given point in time and is determined by employees' evaluation of the factors that define the CultureGRADE®. While QoWL is a continuous variable, it is useful to distinguish among five cultural categories or styles. A given style refers to the group of organizations that obtain similar QoWL scores. Organizations receiving similar QoWL scores have numerous similarities regardless of the industry in which they function and/or the types of products they produce for customers. These five styles are depicted in Exhibit 10. In this exhibit, scores on measures of performance are the dependent or outcome variables and scores on the three QoWL factors are the independent or predictor variables. These cultural styles systematically vary across three broad dimensions:

- Shared meanings

- Internal focus and connectedness

- Externally responsive actions

The level of shared meaning, internal focus, and external responsiveness predict the level of fit-for-purpose results and form the basis for a QoWL Grade. There are five grades, and within a given grade, organizations are similar with respect to four process variables:

- What directs behavior

- What motivates behavior

- Who is empowered to achieve the goals of the organization

- Who is accountable for the consequences of those behaviors

The names of the five cultural styles were chosen to reflect the similarities among organizations receiving a given QoWL grade on each of these processes. The labels reflect variations in the level of efficacy created by those critical decisions made and reinforced by leaders. At the low end of the continuum, employees feel little efficacy or enthusiasm, individually or collectively. While at the upper end, there is a strong sense of efficacy and enthusiasm at both the individual and collective levels.

Grade A. The adaptive culture

Adaptive cultures exhibit a very high degree of:

- Belief in and commitment to the purposes of the organization, trust and confidence in the ability and intentions of senior management, belief the organization cares for and values its employees, and congruence between espoused and practiced values.

- Confidence in and respect for supervisors, collaboration, role and expectation clarity, autonomy and responsibility for achievement, formal and informal performance feedback, and recognition for accomplishment.

- Openness of information flow, a willingness and ability to respond to customers' needs, and commitment to continuous learning and improvement.

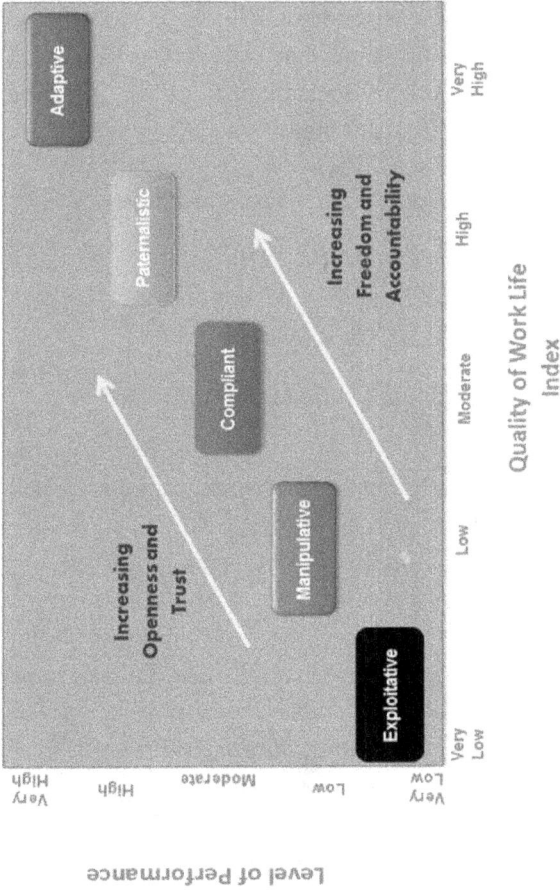

Exhibit 10 *Cultural style is a function of QoWL grade.*
© *Somerset Consulting Group, 2014*

An adaptive organization is one that is able to create sustainable social value. Social Value is defined as creating outputs which serve the greater good and doing so in a way that enables people to engage in healthy pursuits (e.g., jobs, lasting employment, quality of work life, low environmental impact). Adaptation is an observable outcome that reflects the underlying capabilities of the organization. In an adaptive organization, all the parts are:

- Unified around a shared set of meanings and values;

- Internally connected;

- Externally responsive, and;

- Capable of focused action to ensure goodness of fit.

An adaptive system is a seamless, consistent, self-reinforcing alignment of strategy, structure, culture, leadership behaviors, HR policies and processes, and measurements and rewards that enable and promote defined adaptive behaviors. Behaviors produce results and in an adaptive system, values drive behavior and everyone feels a sense of responsibility and accountable for the whole. Adaptation is more likely if leadership:

1. Creates an organizational environment that promotes adaptation and deters behaviors that inhibit adaptation, and;

2. Behaves in ways consistent with that intended environment.

It is this enabling organizational environment that we refer to as an adaptive system. Such systems enable both individual and organizational growth mind-sets, growth behaviors, and the utilization of growth processes.

Grade B. The paternal culture

Paternal cultures exhibit a somewhat high degree of:

- Belief in and commitment to the purposes of the organization, trust and confidence in the ability and intentions of senior management, belief the organization cares for and values its employees, and congruence between espoused and practiced values.

- Confidence in and respect for supervisors, collaboration, role and expectation clarity, autonomy and responsibility for achievement, formal and informal performance feedback, and recognition for accomplishment.

- Openness of information flow, a willingness and ability to respond to customers' needs, and commitment to continuous learning and improvement.

The main differentiating factors of paternal cultures are:

1. Motivation is extrinsic to the individual, and;

2. The individual is seen as being accountable for the success and/or failure of the whole.

Paternal cultures tend to be rules-based, where those wishing to participate in the culture learn the rules and agree to abide by them. The rules are clearly spelled out and workers either embrace them or spend time trying to make things work under the rules. These cultures tend to be management-oriented, with an established managing class and well-entrenched bureaucracy.

Grade C. The compliant culture

Compliant cultures exhibit a moderate degree of:

- Belief in and commitment to the purposes of the organization, trust and confidence in the ability and intentions of senior management, belief the organization cares for and values its employees, and congruence between espoused and practiced values.

- Confidence in and respect for supervisors, collaboration, role and expectation clarity, autonomy and responsibility for achievement, formal and informal performance feedback, and recognition for accomplishment.

- Openness of information flow, a willingness and ability to respond to customers' needs, and commitment to continuous learning and improvement.

The main differentiating factors of compliant cultures are:

1. Priorities are unclear and often in conflict;

2. Motivation is extrinsic to the individual, and;

3. The boss is seen as being accountable for the success and/or failure of the whole.

In compliant cultures, behavior is motivated by the drive to seek pleasure (approval) and to avoid pain (punishment). Motivation is highly extrinsic and produces a space that rewards conformity while inhibiting risk and experimentation.

In so far as its effect on responsibility, compliant cultures try to eliminate the need for individual initiative by relying on well-defined processes and procedures. This is why they are referred to as Control-Order-Prescribe cultures, which try to shape responsibility through the processes of COP. Compliant cultures tend to center responsibility for performance on the supervisor thus compliant cultures are often management-oriented, with an established managing class and a plethora of bureaucratic controls. Finally, compliant cultures are generally incapable of achieving anything near sustainable growth and high performance.

Grade D. The manipulative culture.

Manipulative cultures exhibit a low degree of:

• Belief in and commitment to the purposes of the organization, trust and confidence in the ability and intentions of senior management, belief the organization cares for and values its employees, and congruence between espoused and practiced values.

- Confidence in and respect for supervisors, collaboration, role and expectation clarity, autonomy and responsibility for achievement, formal and informal performance feedback, and recognition for accomplishment.

- Openness of information flow, a willingness and ability to respond to customers' needs, and commitment to continuous learning and improvement.

The main differentiating factors of manipulative cultures are:

1. Priorities are unclear and often in conflict,

2. Motivation is extrinsic to the individual and is based on fear and avoidance of negative consequences, and

3. The boss is seen as being accountable for the success and/or failure of the whole.

In manipulative cultures, rarely does anyone question the way things are done for fear of negative consequences. It is this fear of negative consequences that drives behavior. Employees do what they think they are expected to do and behavior is motivated by the drive to avoid pain (punishment). Motivation is highly extrinsic, produces a space that rewards conformity and punishes deviation from the existing norms.

In so far as its effect on responsibility, manipulative cultures try to eliminate the need for individual initiative by relying on well-defined processes and procedures. These cultures are reactive and seek to shape responsibility through the processes of control, order, and prescribe. However, they differ from compliant cultures in that manipulative cultures react to the environment more than they plan for it. Responsibilities for performance tends to center on one-up managers and supervisors are relegated to the role of inducing compliance. Finally, manipulative cultures are generally incapable of achieving anything near sustainable growth and high performance.

Grade F. The exploitative culture

Exploitative cultures exhibit a very low degree of:

- Belief in and commitment to the purposes of the organization, trust and confidence in the ability and intentions of senior management, belief the organization cares for and values its employees, and congruence between espoused and practiced values.

- Confidence in and respect for supervisors, trust in peers, collaboration, role and expectation clarity, autonomy and responsibility for achievement, formal and informal performance feedback, and recognition for accomplishment.

- Openness of information flow, a willingness and ability to respond to customers' needs, and commitment to continuous learning and improvement.

Exploitative cultures are pathological. The main differentiating factors of exploitative cultures are:

1. There is no compelling vision and priorities are unclear and often in conflict.

2. Motivation is extrinsic to the individual and is based on fear and the need to survive, and

3. The boss is seen as being accountable for the success and/or failure of the whole.

In exploitative cultures, rarely does anyone ever question the way things are done for fear of negative consequences. It is this fear of negative consequences that drives behavior. Employees do enough to get by and to make the system work and their behavior is motivated by the drive to avoid pain (punishment). Motivation is highly extrinsic and random and produces a space that rewards expediency. Behavior is motivated by fear and not getting caught.

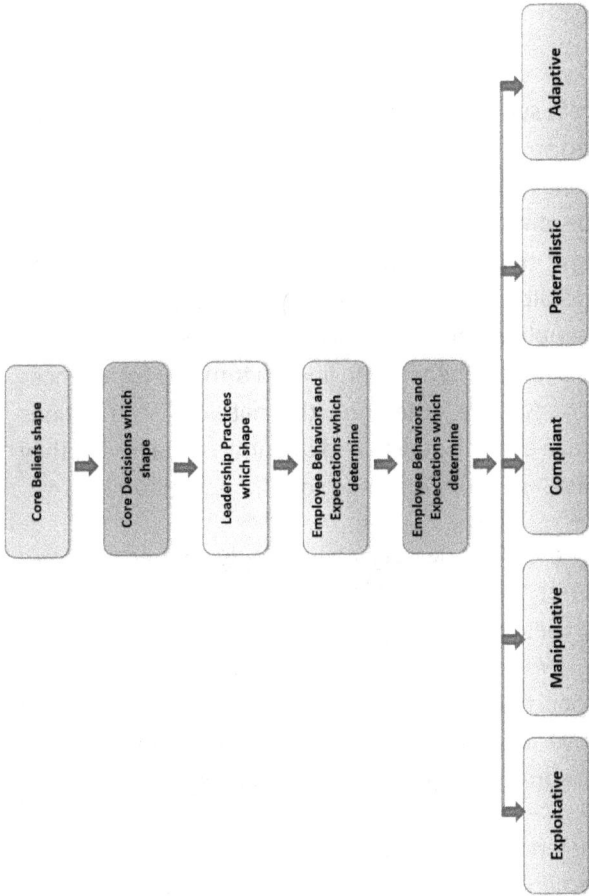

Exhibit 11 *From decision to reality—The evolution of cultural.*
© *Somerset Consulting Group, 2014*

Core Beliefs shape → Core Decisions which shape → Leadership Practices which shape → Employee Behaviors and Expectations which determine → Employee Behaviors and Expectations which determine

Exploitative | Manipulative | Compliant | Paternalistic | Adaptive

Conclusions

*We shall not cease from exploration and the end of all
our exploring will be to arrive where we started and
know the place for the first time.*

T.S. Elliot, East Coker

The data summarized above show that adaptive cultures
outperform cultures that are less adaptive. The data
also showed this evolution is very systematic and
can be predicted as illustrated in Exhibit 11. Culture
emerges out of a shared meaning about value, reality,
and possibility. These shape the kinds of experiences
that people will have as well as the routines and rewards
needed to keep behavior in line. In turn, the experiences
employees have determine the nature of the actions
they take and the quality and quantity of the results they
produce. These in turn lead to different cultural styles.

Thus we can argue very convincingly that an adaptive
organization is one which has the ability to engage
employees in the on-going creation of meaning and
value. What our research has shown is that the higher
an organization scores on this set of variables the more
adaptive it is. In summary, a learning culture shows a
high degree of alignment in three areas:

- Senior leader's alignment with the
  organization's vision mission and values
  (Shared meaning);

- System and process alignment with
  organizational vision, mission, and goals
  (Internal Connectedness and Focus);

- Employee alignment with the market demands (External Responsiveness).

Interestingly a recent study in the McKinsey Quarterly (2011) showed that senior leaders regularly kill meaning by falling into four traps:

- Constantly shifting priorities;
- Settling for mediocrity;
- Disorganization, and;
- Misbegotten goals.

In one study, we surveyed the customers of an oil exploration and production company. Customers who rated the company high on the first-order predictive variables (Exhibit 5) were highly satisfied with their relationship with the company, more likely to remain loyal to the company, and less likely to have tried competitor's products and services. The opposite was true for those customers who evaluated the company negatively on these same variables. Furthermore, a majority of customers in this study rated the company negatively. Leadership did not act on the data presented and the company has since gone out of business.

In another study we conducted of utility companies, the companies which were perceived to be adaptable obtained significantly higher satisfaction and loyalty scores from customers. Significantly, the customers of the two companies with the lowest adaptability were exploring the possibility of the co-generation of electricity, a move which threatened both their base rate, and ultimately, their survival.

To summarize the available data, it seems as if the following conclusions about the relationship between culture and performance can be drawn:

- Corporate cultures have a significant and measurable impact on long term economic performance and culture will become even more important in the future as an issue determining the success or failure of organizations;

- Frequently, cultures act to inhibit long term financial performance;

- Those cultures which are adaptive are more likely to achieve higher levels of economic performance than those which are not;

- Cultures can be made more performance enhancing if leadership is willing to take a long-term view.

Part 3: Crystals learn

They always say that time changes things,
But actually, you have to change them yourself.

Andy Warhol
The Philosophy of Andy Warhol:
From A to B and Back Again

The third part of this paper is concerned with detailing a model and practical action steps for improvement centered on creating open, adaptive cultures. The metaphor of crystals learning is an apt one. Crystals don't just learn willy-nilly, and the conditions have to be right. For example, crystalline growth is dependent on temperature, the relative concentration of the medium in which the crystal is located and time.

Crystalline structures in nature are structures which, at a very low level, have learned successful replicating strategies. These structures start with a single microscopic seed around which others cluster, in orderly fashion. It takes about half as long for the second crystal to form (or 'learn") as it did the first. In addition, the crystal continues to grow in an exponential fashion.

In a physical sense, the crystal is the mirror of its interlocking molecules. Regardless of their substance, the angles between any two crystals are always identical. Crystals have no genetic imprint, yet they always grow according to a certain set of criteria which shape their physical presence and internal structure. The crystal does not know those criteria, any more than as individuals we can know the genetic structure which has created us as unique persons. But the crystal has learned some things in the same way that organizations can learn: by examining structure, past behaviors, arrangement of sub-systems, and adaptation to the environment—all of these are empirically observable in crystalline growth.

So, in a sense if we said that crystals are not alive, we'd also be saying that our organizations are not alive; that they possess no force, no impetus to action or lassitude, no defenses, no ways of understanding their environment, no inkling of the complex inter-weaving of systems in which they exist. But we know this isn't true. Cultures can learn and grow if the external and internal conditions are right. The exploration of these conditions is the focus of Part 3.

Myths versus practical assumptions about cultural adaptation

Before getting to the issue of culture change and adaptation, let's address some of the myths that surround the topic:

Myth 1: Culture is a fix for any problem you might face

As we learned in Part 2, culture is relatively stable, enduring and serves to inhibit change. So, in the long term, culture is not a fix for anything. What brings about a change in culture is a change in behavior and what normally brings about a change in behavior is a change in some aspect of practice or strategy.

Myth 2: Culture and strategy have nothing to do with one another

The truth is that in terms of execution, culture (behavioral patterns deeply ingrained in people) and strategy (an idea about how to compete effectively) are inseparable. For a strategy to succeed quickly, it must take advantage of this cultural inertia. It must channel people's energies into actions they are comfortable taking. A strategy that asks people to do something unnatural or totally foreign is doomed to a slow death.

Myth 3: Cultural change can be managed

Cultures change only when they need to and are ready to change. Actually, they change when their collective intelligence recognizes that the world has changed and that the culture should adapt in order for the business to survive. When it comes right down to it, culture is about surviving and thriving. But profound cultural change is a messy and painful process. It takes quite a while for individuals to get their heads and hearts attuned to the fact that the world has changed and they must change, too.

Myth 4: Top-level leadership is the key to instilling a strong corporate culture

Leadership has a lot to do with building a cohesive culture, but it is not just the leadership that allows the company to carry out its economic mission successfully. It is leadership that seeks to shape a working environment that people at all levels can identify with, that encourages leadership from everyone, that is not afraid to stand for something, that cares about a myriad of details that make the company work, that strives to generate universal pride in what the company accomplishes, not just how people are enriched by its economic activities. It is leadership that all good managers should exercise if they take themselves and their responsibilities seriously.

Myth 5: People hang onto a culture they know even when it is no longer relevant

People change willingly when they are convinced of the need for change and when they agree with the direction decision makers are proposing they go.

Myth 6: Cultures resist change

Cultures in companies are the living, breathing manifestation of the most deeply held desires of people to do what's right and get ahead. Because of this, cultures thrive on change just as much as they champion tradition. New and better ways of doing things; new strategies in the marketplace; new offices, plants, and decorations: cultures love them as signs that all is right with the world. Cultures are always adapting to the changes around them. Failure to adapt would be threatening since it would be seen as a sign that the culture was falling behind. Where cultures resist is when long-standing core values or widely accepted rituals or practices are endangered. And they resist such changes because the force of history is behind them: Is this proposed change really what's needed to move on?

Myth 7: Culture is not for everyone

Like it or not, one is immersed in a culture at work. Whether you think it is important or not, cultural mores and norms dictate much of what one does on a day-to-day basis and determines how you think. Whether you conform or not is not a matter of choice-unless you want to leave. The culture of a company seeps into your pores and shapes your identity.

Practical and positive assumptions about adaptation

We know for a fact that cultures have the ability to learn and adapt to new circumstances; they are always responding to the environment and redefining themselves. Here are some beliefs or expectations that are typical of adaptive cultures:

Expectation 1: Proactivity

An adaptive culture must assume that the appropriate way for humans to behave in relationship to their environment is to be proactive problem solvers and learners. If the culture is built on fatalistic assumptions of passive acceptance, learning will become more and more difficult as the rate of change in the environment increases. Learning-oriented leadership must portray confidence that active problem solving leads to learning and adaptation, and in doing so, set an appropriate example for other members of the organization.

Expectation 2: Commitment to learning how to learn

Cultures that adapt have gained the ability to learn. They have developed the means to gather and use data from the external world. Furthermore, they have developed the means to gather and use information from their internal environments to ensure that a response to an external change does not disrupt or lead to the disruption of internal connectedness. Adaptive organizations have learned how to recognize and be open to feedback, generate new responses, and to learn from experimentation and reflection.

Expectation 3: Positive assumptions about human nature

An organization cannot adapt unless it is willing to give people the freedom and responsibility for generating new responses. This requires believing that people are trustworthy and capable of ethical behavior. It also assumes that people are able to learn and grow, and will if given the chance.

Expectation 4: Belief the environment can be managed

Recall that expectations are self-fulfilling. This means that an adaptive or learning culture must contain in its DNA a gene that reflects the shared assumption that the environment is to some degree manageable. That is to say, there can be no doubt that change is possible and can be achieved through determined effort.

Expectation 5: Commitment to truth through pragmatism and inquiry

A learning culture must contain the shared assumption that solutions derive from a deep belief in inquiry and a pragmatic search for "truth." Hence there is no one way to skin the cat but many ways in each of which inheres a range of benefits as well as negative consequences. Adaptive organizations approach change with "eyes wide open."

Expectation 6: Positive orientation toward the future

An optimal time orientation for learning appears to be somewhere between the far future and the near future. We must think far enough ahead to be able to assess the systemic consequences of different courses of action, but we must also think in terms of the near future to assess whether or not our solutions are working. This may be thought of as double-loop learning (Argyris & Schön, 1978).

Expectation 7: Commitment to open, task relevant communications

An adaptive culture must be built on the assumption that communication and information are central to organizational well-being and must therefore create a multichannel communication system that allows everyone to connect to everyone else.

Expectation 8: Commitment to cultural diversity

The more turbulent the environment, the more likely it is that the organization with the requisite variety of agents and resources will be better able to cope with unpredicted events (Ashby, 1968; Salem, 2008). Therefore, the learning leader should stimulate diversity and promulgate the assumption that diversity is desirable at the individual and subgroup levels. Such diversity will inevitably create subcultures, and those subcultures will eventually be a necessary resource for learning and innovation.

Expectation 9: Commitment to systematic thinking

As the world becomes more complex and interdependent, the ability to think systemically, analyze fields of forces and understand their joint causal effects on each other, and abandon simple linear causal logic in favor of complex mental models will become more critical to learning. The learning leader must believe that the world is intrinsically complex, nonlinear, interconnected, and "over determined" in the sense that most things have multiple causation (Owen & Dietz, 2012).

Expectation 10: Belief that cultural analysis is a valid set of lenses.

In an adaptive culture, leaders and members believe that analyzing and reflecting on their culture is a necessary part of the learning process. Cultural analysis reveals important mechanisms by which groups and organizations function in completing their tasks. Without cultural analysis, it is difficult to understand how groups are created, how they become organizations, and how they evolve throughout their existence.

Cultural change

Before we focus on the notion of culture change let's make a distinction between adaptation and learning. Adaptation is the ability to make changes that improve the organization's survivability and ability to perform at a sustainable level. Change, is the outcome of a process that includes both organizational learning and adaptation and learning is one of the processes which makes adaptation possible. There are three aspects to adaptation: the existence of a need to change, the ability to learn which changes are likely to enhance sustainability, and the ability to respond to that need through adaptation. Which is the more pragmatic construct—the notion of adaptation or learning? In our view it is more fruitful to think about the process of adaptation and consider the ability to learn as one of the competencies required to achieve the aim of adaptation—survival and sustainability.

The task of changing your organization's culture can seem a bit overwhelming to even the best and most committed leaders. There are many reasons for this, however, the most common reason is that most organizations, and therefore, the individual leaders in the organization, have not clearly defined what culture means and looks like for them. So the first hurdle is to start thinking about what your culture would look like if it were a high performing, sustainable culture, a culture in which production, cost, quality, and caring are all integral parts of how you define success as a business. Performance excellence is the goal. To the enlightened manager and leader, performance includes operational excellence, environmental excellence as well as injury free operations.

The role of leaders in cultural change: leaders as pathfinders

Cultural change leaders are like pathfinders (Read, 2010). They are able to map out a journey to take their organization from where it is now to a new level of thinking and performance. The metaphor of a pathfinder allows us to understand what this process looks like and what steps are required to achieve success. A pathfinder is someone who discovers a way; their role is to clarify and communicate the path so that the rest of the organization can move to a new location.

Pathfinders are expert navigators who are skilled and competent in the science of using a map and compass; however, they understand that navigation is also an art. Because a map is not the actual terrain but only a representation of the terrain, it is not completely accurate. Likewise, compasses can be affected by magnetic interference and are also not completely accurate. Pathfinders are able to apply the science and map out the journey they need to take. Plus they are also able to apply the art of navigation and interpret (make sense of) what they find on the ground.

In the same way a Pathfinder uses a map and a compass to navigate the path, an organizational leader maps out the steps needed to change or transform culture. As a role of leadership, a Pathfinder needs to know how to use a map and compass; however, these tools alone are not enough. A leader also needs to know the mechanics of using a map (how to calculate a bearing, read distances, understand the terrain, etc.); and how to work out a route from where the organization is currently to where it wants to be. Finally, leaders must be able to pinpoint where their organization now rests. Knowing the current state of the organization enables leadership to map out the best possible course for their journey. You can determine how far the organization needs to move to become effective at meeting its goals and what the journey will require in terms of resources, time, effort etc.

Transforming the culture

Change leaders must use a path-mapping process. In the same way that a pathfinder creates a path from here to there, an organization's leaders create the path to transform the culture from where it is to where it needs to be—a sustainably high performing culture. The *CARES* framework provides such a map (Winters, Owen, Read, and Ritchie, 2010). Based on the Influence Model presented in Exhibit 4, CARES describes a process for creating an achievement focused, relationship-based endeavor sustainably. *CARES* is like the map and the compass; it allows you to map out the journey needed to create a culture change. The *CARES* framework allows the leader to map out the journey to sustainable performance excellence (see Exhibit 12).

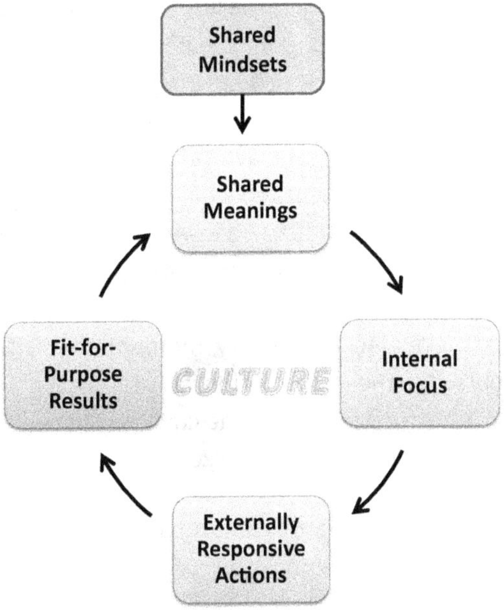

Exhibit 12 *The CARES framework*

Leaders transform culture by the *meanings* and *experiences* they create to focus the organization on what is vitally important—steering their culture to a mapped out vision. They do the things they need to do, in order to make the difference they believe is needed. They are in the game (in fact they are "All in"), they choose to be responsible and they hold themselves accountable. Of course they make these choices within the constraints and limitations of their current thinking and their paradigms. The most effective leaders are aware of this and are constantly exploring what else they could do that would make a difference. A good question every leader should ask themselves regularly is, "What is shaping my ability to contribute to the organization's culture?"

This question is a very powerful one for a leader. An organization's culture gets created by design or by default. If leaders are not steering their organization then the organization is like a ship without a rudder—it will go whatever way the wind and currents take it. In other words, the organization will be responding to the environment in a way that allows the organization to cope with environmental change as opposed to fitting better with environmental change. Organizational leaders make decisions every moment that influence the course of the organization as well as the change leader's efforts to transform the culture. There are three critical outcomes to any organizational decision:

1.   What meanings am I creating and communicating,

2. What experiences am I creating to ensure we are focused on doing the right things, and

3. What actions are we taking to ensure we are producing fit-for-purpose results?

Answers to these questions determine the choices that are made. They directly shape the effectiveness of an organization's leadership in transforming their organization. If leadership is not clear or if leadership is not consciously making decisions with a plan and a clear intent in mind then the organization will never end up moving/changing in a coherent/sustainable manner.

Effective transformational leaders achieve success by creating and managing the four building blocks: shared meanings, meaning focused experiences, responsive actions, and fit-for-purpose results. As leader you MUST know what you want to achieve (Shared Meaning) and be able to map out the path to get there (Focus and Action). If you do not know where you want to go then, it doesn't matter which path you take. This is summed up well in the saying: "The reason most people don't achieve their goals is that they never really set them in the first place."

Relationship is the foundation of achievement

Everything is created twice—first in our minds and then in the physical world. To create our vision we must truly understand what we are trying to achieve (Shared Meaning). We need to be able to picture the goal and what it looks like once we have achieved it. First we get clear mental picture about what we value, what we want and what it looks like—this is what we mean by Shared Meaning; it's our relationship to our vision—sustainable performance excellence. Only then do we apply the science of achievement to map out what we need to Focus on and the Actions we need to take to achieve that goal.

Our meanings define our relationship to the culture

Meanings are the attributions we make based on what we believe to be possible, important, and real. These meanings aren't obvious to others; rather we have to make them so via the relationships we create at any given moment.

1. Our relationship to concepts and things, e.g. our relationship to the subject of our vision. The deeper our relationship the greater the level of understanding and meaning we create.

2. Our relationship to others. Our ability to build and maintain trust and connectedness with other people.

3. Our relationship with ourselves. Of the three relationships, this is the most powerful and significant for a leader. If we do not know our values and why these are important to us we will not make consistent or enlightened decisions. Our relationship with ourselves is about being clear on our character and on being comfortable with the 'man in the mirror'.

Focused experiences

What we focus on is determined by our beliefs (values and expectations). Our beliefs determine what things <u>Mean</u> to us: is it good or bad, possible or not possible, important or not important, urgent or not urgent? Based on our answers to these questions we make choices regarding what we <u>Focus</u> on. Our beliefs influence our understanding and thus shape our focus, especially in terms of how to best achieve what we want. George Clason (1926) captured this in his book *The Richest Man in Babylon*, when he said:

> *Our acts can be no wiser than our thoughts, our thinking no wiser than our understanding.*

The best leaders maintain their focus (and the focus of their organization) with a laser like intensity on what they know is right and needs to happen. Changing one's focus can change what one sees, including opportunities or solutions that may not have been apparent from one's previous point of view. Others see that, people know what leaders care about, what is important to an organization's leadership—that is focus.

Externally responsive actions

While the experiences leaders create provide the container, the actions employees are expected to take ultimately determine the results produced. Our actions determine what we create, however; they also determine sustainability. The challenge for the leader is to empower people to take actions that not only produce results but also enable employees to create small wins every day.

External responsiveness means having the ability to take in, assimilate, and use valid data for the purpose of providing the products and services customers want and expect. There are three things leaders must do to enable employees to take externally responsive actions. Leaders must provide:

1. Role clarity—to what extent does each person understand how his or her role and to what extent does this role definition include the freedom to learn and grow?

2. Expectation clarity—what level of performance is expected of each person and does she or he have the power and freedom to accomplish this level of excellence in a timely, cost effective manner?

3. Information clarity—to what extent does each person have the data she or he requires to assess what to do, how well it is being done, and how to make it better.

It might be said that the hallmark of a healthy organization is the degree to which it is able to create, maintain, and reinforce clarity of meaning. This is because each of us gets to decide what something means to us—e.g. is it important, what emotion we bring to a situation, and what this means for the present and the future. What we think and how we feel about something is what determines the quality of our inner work life, how motivated we are, and the level of energy we will devote to our endeavor.

A fit-for-purpose endeavour

A hallmark of the Adaptive culture is that these organizations have learned how to learn. They use this ability to continually adjust the results they produce to fit the demands of the environment in which they have chosen to compete. The culture style ladder, in fact, provides a framework for how capable you are at doing just this. Where your organization sits on the ladder is an indication of the current meanings you have given to things and have reinforced through the experiences you have created and the actions you have taken.

A company that is at the Manipulative level will have attached different meanings to events than a company at the Adaptive end. For example Manipulative companies respond to a high potential near miss incident (HPI) as if they "got lucky" and "dodged a bullet" and as a result they don't take any action, and they don't learn how to eliminate the risk. On the other hand, a company at the Adaptive end attributes an entirely different set of meanings to an HPI. They think of them in the same way as an actual hit and they act accordingly to eliminate any future risk. This is because companies at the top of the ladder attach different meanings to the concept of risk. In addition, they manage risk in an entirely different manner than those lower down the cultural style ladder. They understand the risk and are clear about the disciplines they must maintain to safely work with that level of risk.

In any case, a company's past does not have to be its future—you have choices and understanding the culture Ladder allows us to understand what meanings are guiding that endeavor at both an organizational and individual level. It is the attributions of meaning at the executive level that shape and influence the leadership practices in the organization. Leadership practices create the experiences that people have and these fall along a continuum (Exhibit 13).

1. Fear based leadership practices;

2. Incentive/Disincentive (carrot and stick) based leadership practices, and;

3. Values based leadership practices.

These leadership practices work on different human needs as shown in Exhibit 13. Fear based leadership practices target the lowest level of our thinking and are what we refer to as Motivation 1.0 practices. A typical example of leadership at this level is threatening to sack someone if they don't do what you want. Of course it's generally more subtle than that but we all know what this leadership climate feels like. You start to get a feeling in your gut that if you speak up you may not be around too long. This type of leadership motivates people to not get caught. It leads to game playing and many people becoming very good at playing the game. People are not focused on following the rules; instead they are just focused on not getting caught. Sounds like a traffic cop doesn't it? The product of leadership is engaged people and teams. So in reality, we shouldn't actually be calling this leadership because it creates zero engagement; in fact it leads to disengagement.

Carrot and stick based leadership is a transactional based leadership practice—if you do this then you'll get

Creating Sustainable Performance

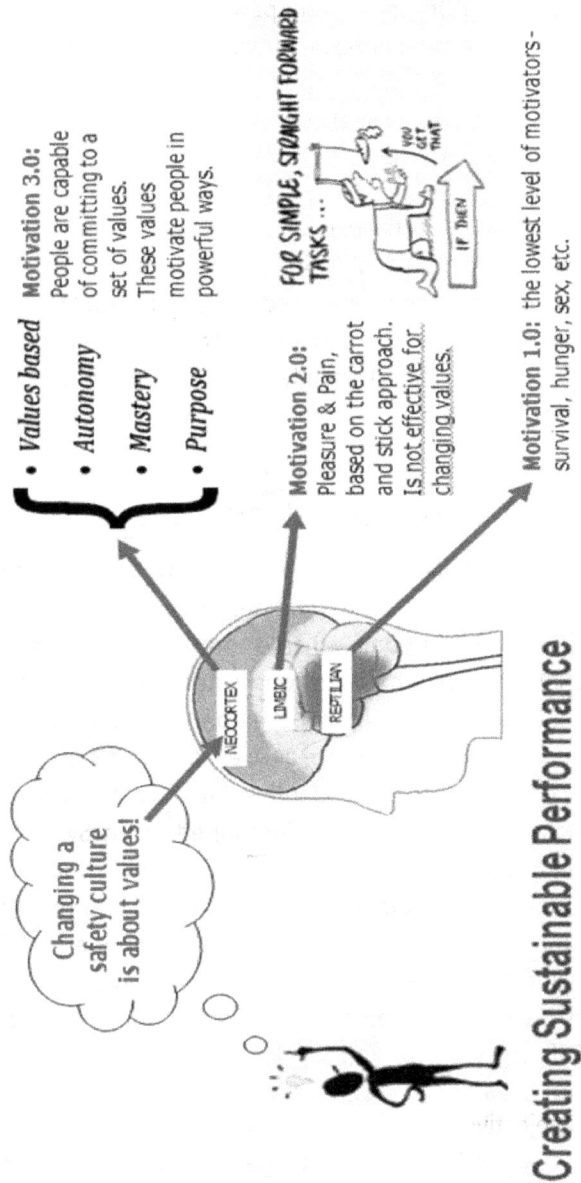

Changing a safety culture is about values!

- *Values based*
- *Autonomy*
- *Mastery*
- *Purpose*

Motivation 3.0:
People are capable of committing to a set of values. These values motivate people in powerful ways.

FOR SIMPLE, STRAIGHT FORWARD TASKS ...

YOU GET THAT

IF THEN

Motivation 2.0:
Pleasure & Pain, based on the carrot and stick approach. Is not effective for changing values.

Motivation 1.0: the lowest level of motivators - survival, hunger, sex, etc.

NEOCORTEX
LIMBIC
REPTILIAN

Exhibit 13 *The three kinds of motivation*

that. It is great for motivating people to produce more widgets on a production line (at least for a short while). It gets compliance but it stifles creativity. It does very little for engagement and commitment. We call these Motivation 2.0 practices.

Values based leadership is about leadership practices that are focused on Motivation 3.0 practices. To create sustainable excellence we need engaged people who are committed to creating excellence for themselves and their workmates. This only happens at the Adaptive level of leadership.

The integrated open systems model

Building an adaptive culture starts by creating an achievement focused, and relationship based endeavor. It is aligned around shared meanings, focused experiences, external responsiveness and fit-for-purpose results. We have developed a model which is useful, both heuristically as well as practically, for understanding what an adaptive organization is like. We call this (Exhibit 14) the Integrated Open Systems Model (Dietz & Owen, 2012). The first thing to note is that there are two dimensions of organizational reality: the subjective and the objective which are understood at the individual and the organizational levels. This produces four quadrants each of which obeys its own set of laws.

The objective side of the organization comprises individual skills and knowledge and the processes and systems designed to integrate them so they

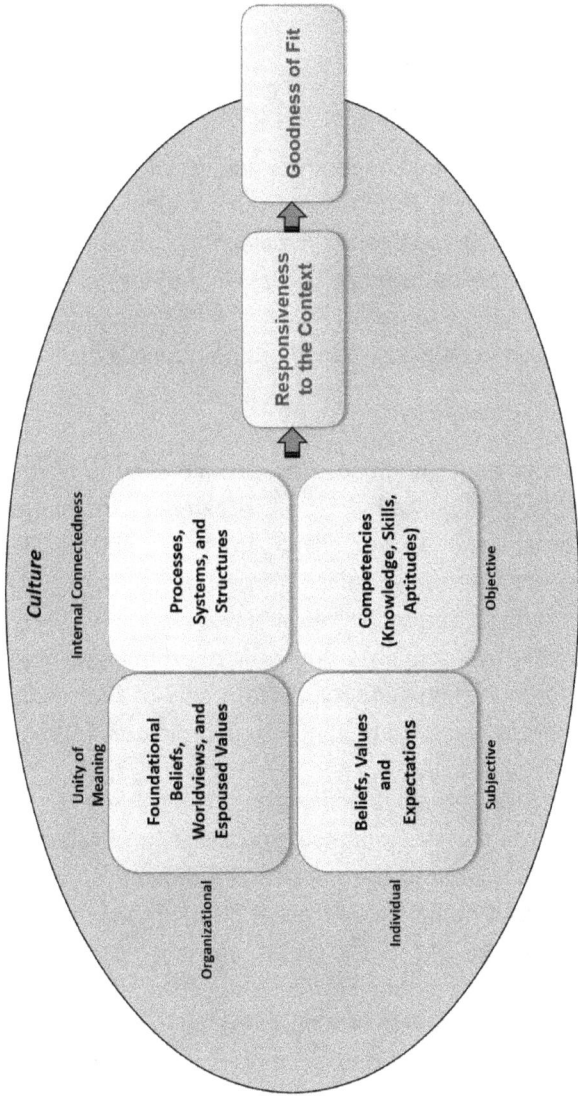

Exhibit 14 *The integrated open systems model*

work in concert towards achieving the purpose of the organization. The objective side of the organization is observable and measurable and tends to follow the rules of engineering.

Behavior quadrant

The lower right hand quadrant is the domain of behavior, and the view of the individual from the exterior. It is all the things that you see the individual doing or working with, and the knowledge and skills they bring to the job. Improvements in this area come from working with individuals to influence and align their behavior.

Systems quadrant

The upper right quadrant is the domain of systems, the view of the organization from the exterior. It includes the organizational and management structures and the systems required to align work activities (e.g., IT, measurement, correction, rewards and recognition, etc.). Change in this domain is driven by good management. Most organizations understand this side of the model, but they don't integrate it into the big picture.

The left side is the subjective view. The subjective side of the organization comprises the beliefs and values of individual members, their hopes and dreams, and their expectations about themselves and the organization. It also comprises the vision, values, and mind-sets of the organization as a whole. The subjective side cannot be directly observed but must be inferred from the perceptions and actions of members and decisions and actions of the organization. Many organizations struggle to understand this side of the model. For these organizations, this lack of understanding prevents them from seeing the big picture.

Intention quadrant

The lower left quadrant is the domain of the intention, and the view from the interior of the individual (consciousness). It includes the values, commitment and the intentions the individual brings to all situations. Improvements in this area come from working with individuals, through leadership, and improving the sense of belonging. Change in this area is typically perceived as difficult and requiring time. In reality a change in intention can happen in an instance.

Cultural perspective quadrant

The upper left quadrant is the domain of shared perceptions, the view from the interior of the group. It includes the actual or perceived shared values, norms and standards of the group. It is here we find the ethics, morale and sense of justice that is commonly held by the group. Positive change in this domain has its genesis in leadership.

Access to the objective side is via monologue. It is the documents, policies, procedures, training manuals, observable behaviors, etc., that tell people how things are to be done, what's important, and how to act, operate and behave. Access to the right side is via dialogue. In real life, it may contradict the monologue and tell people how things are really done around here. While the subjective aspect of this side cannot be observed or measured directly the dialogue can be observed, recorded and reported upon. However what the dialogue means, what motivated it and what the expected or anticipated outcomes of the dialogue were also sits in the subjective realm. The subjective side of the model is the focus of leadership. Effective leadership always has a strong grounding in dialogue and is typically focused on a view or vision of the future. What the possibility is for that future is expressed by leaders in their dialogue and their conversations in the organization (Shared Meaning).

Each dimension follows its unique set of laws and principles: the subjective dimension obeys the laws and principles of individual and group psychology and complexity theory. Whereas the objective dimension follows the laws governing the physical universe.

Integrating meaning, focus, action and results—Key to transformation

Integrating meanings, focus, action, and results is the key to cultural transformation and this is the primary task of leadership as adaptation requires their ongoing involvement and commitment (Mink, Mink, Downes & Owen, 1994; Dietz & Mink, 2005; Owen & Dietz, 2012). The extent to which organizations are able to handle the complexity of continually integrating shared meaning, internal connectedness, and responsiveness to valid information within a specific context is the extent to which they will be successful at adapting (Dietz & Mink, 2005). In fact, Jacques and Clement (1991) maintain that:

...handling complexity is at the heart of the competence to deal with problems. How well or how badly managers handle their problems is in turn at the heart of not only the way in which they are regarded by their subordinates but also the strength of their leadership. (page 9)

The principle of shared meaning

Leaders have two important jobs: one is to develop a relevant strategy and two is to create experiences that imbue the strategy with personal meaning. Every employee has an inner work life—the constant flow of thoughts emotions, perceptions, and motivations that constitute a person's reactions to the events of the work day. Beyond affecting the employee's well-being, inner work life also affects the bottom line (Amabile & Kramer, 2011). Adaptive cultures have a high degree of 'unity' or shared meaning around vision, values, and goals, operations and processes, and the value of people.

Shared meaning is not expressed in isolation but is forged out of the need and desire to adapt to and successfully cope with the external environment, and is an important determinant of the ability of an organization to learn and grow. The concept of meaning refers to "the symbolic significance members attach to the organization's vision, mission, values, and desired results" (Dictionary. com). In the truest sense shared meanings provides the framework for all organizational behavior. When an organization is able to align members, processes, systems, and aspirations around a sense of shared meanings about what is important and worthwhile, it enables its members to fulfill important psychological needs for purpose, achievement, belonging and respect, and this, in turn, creates high levels of commitment and motivation.

Shared meaning is not a given, however, for it requires members to develop and assign a particular and specific meaning to information and to develop a shared grasp of its significance or its implications for their own as well as others' behavior. At issue then is how organizations go about creating shared meaning, especially during times of needed change and how they 'live' this meaning on a day to day basis. Change is more likely to be successful if leadership: 1) creates an organizational environment that promotes change behavior and deters behaviors that inhibit change and 2) behaves in ways consistent with that intended environment. This is because human systems are highly sensitive to inconsistencies between espoused meanings and actions (Hess, 2012; Amabile & Kramer, 2011).

The principle of internal connectedness

To be connected is to operate as a unified system in order to create and reinforce the shared meanings of the organization. The principle of internal connectedness refers to the way individual capabilities, expectations and aspirations, systems and processes, and shared collective practices are brought together so as to realize the purposes of the organization. The aim of the integration of people, processes, and working relationships is to create a space in which people can engage in purpose-relevant experiences. Purpose-relevant experiences are those that are clearly interpretable as being relevant and contributing to mission accomplishment. In a sense, the shared meanings enable the creation of a space in which people can engage in experiences and take actions that are consistent with the nature of the desired change.

The principle of external responsiveness

Adaptive organizations have learned how to gather and effectively use data from their environment. External responsiveness is the term we use to describe this capacity. The principle of external responsiveness refers to the ability of the organization, its units, and its individuals to assimilate (gather) and accommodate (use) information to increase the goodness of fit with the environment. Assimilation and accommodation are the two complementary processes of adaptation, through which awareness of the outside world is internalized. Although one may predominate at any one moment, they are inseparable and exist in a dialectical relationship.

Assimilation is the capacity to gather and take in information for the purpose of understanding the ever-changing environment. In doing this organizations must exhibit intelligence in choosing which information is relevant to making decisions about how to respond to that environment. Second, organizations must be capable of accommodating the data, that is, it must be capable of 'self-extension' or adaptation. Self-extension is an alteration or adjustment in the capabilities, systems, individual expectations and/or culture by which the organization improves its condition in relationship to its environment. It is also a coordinated change in behavior of individuals and groups in response to new or modified surroundings.

In assimilation, what is perceived in the outside world is incorporated into the internal world, without changing the structure of that internal world, but potentially at the cost of "squeezing" the external perceptions to fit—hence pigeon-holing, siloing and stereotyping. In accommodation, the internal world has to accommodate itself to the evidence with which it is confronted and thus adapt to it, which can be a more difficult and painful process. External responsiveness means the organization learns the ability to discriminate information which is fit for purpose from that which is not, which in this case refers to the degree that new data enhances the ability of organization to realize its purposes. Information is valid when it is relevant to and efficacious for decision making and action.

External responsiveness, then, is the organization's collective ability gather and use information in order to respond to the forces and changes in the internal and external environment. In many organizations, however, external responsiveness is often diminished because decision makers tend to operate from the objective side of the model, both by virtue of their training, and also, by virtue of the fact that information in the subjective side of the model is often seen as threatening. Organizations struggle to understand the subjective side of the model because the level of trust and openness is insufficient to enable a dialog about that which is perceived to be threatening.

As with internal connectedness, to be externally responsive, organizations must address both objective and subjective issues. For example, when introducing new technology, not only must costs verses benefits be considered but also the concerns of those who are expected to use the new technology must be addressed. While the former are obvious, less obvious is the impact of consumer and employee concerns on the adoption of a new technology. Concerns are the feelings people have about the technology and shift in focus over time from awareness to personal to task to innovation and improvement. Unless and until organizations understand these subjective concerns customers and employees may resist even what appears to be a highly beneficial change. The reality is that organizations often overlook these subjective issues and this is because of decision maker's tendency to view organizations in mechanical or engineering terms (objective dimension) and to ignore the more subjective, yet powerful forces of human beliefs and values and shared collective practices (subjective dimension).

The principle of goodness of fit in relation to a given context

Every decision made and action taken by an organization occurs within a context or setting. Context is defined as, "the set of circumstances or facts that surround a particular event or situation" (Dictionary.com). An organization's context refers to the immediate and extended environment in which its purposes are to be achieved and, to be successful, decisions and actions must be fit for purpose in the context in which they function. At issue in achieving goodness of fit is the degree to which the subjective and the objective elements of organizational reality are integrated and aligned. Each of these qualities is present to a greater or lesser extent at three levels—the individual, group, and organization.

A process for cultural transformation

Readiness for change

To move the organization from where it is to where it needs to be, change leaders have to make sure four pre-requisite conditions exist and are widely understood:

1. Power to influence adaptation

As we have said, transformation is the process of creating an achievement focused relationship based endeavor. Adaptation thus requires influence and power and it is usually those in leadership roles who have the power to influence. Bennis (1984) found that the most effective leaders of adaptation did four things extremely well:

- They managed attention and kept the organization and everyone in it focused on the vision mission and values of the organization.

- They managed meaning by consistently communicating in word and deed the importance of the vision to the long term health of the organization to all those constituencies whose cooperation are needed so as to influence the creation of teams and coalitions that are crucial to the success of the organization.

- They managed trust by building a culture in which truth and honesty are valued, where keeping your commitments is the norm, and where it is safe to take risks and innovate.

- They effectively managed themselves by understanding themselves and using their skills in the most effective manner possible.

Kotter and Heskett (1992) extended the findings of Bennis and found that leaders must also be able to see how to transform the organization. Kotter and Heskett refer to this as the ability to see the organization from an outsider's perspective. When you are part of the culture you cannot see the culture. Thus, in some fashion, change leaders are able to get outside the system, so to speak, and see it as it is—not as they expect it to be. This capacity enables employees to see what must be done in order to bring about changes. In every case of successful change reported, leaders were able to win over the people who could help them the most in their efforts to transform the culture.

2. The motivation for change

People don't change unless they see the need to do so and effective change leaders make sure people understand the gap between where the organization needs to be and the current state of the organization. They create a crisis of opportunity. For example, to provide the energy for its cultural change one oil company we work with drove home the `painful fact that they were not even average when it came to transforming crude oil into gasoline ready for sale at the pump. Leadership also stressed the fact that at the current level of efficiencies, the company could not long survive as a competitive force in the oil industry. The most successful change leaders create an intensely felt perceived need for change.

3. Strategically appropriate direction for the change

It is not enough to provide energy. Leaders must also position their organizations in a direction that makes economic sense. The most effective leaders spend a lot of time talking about and educating people about the vision of what changes are needed. They paint a clear and meaningful picture of the ideal as compared to the current reality.

4. The structure within which change can take place

Finally, successful leaders create a structure in which hundreds or even thousands of small, change initiatives could occur simultaneously. And in looking for opportunities, these leaders focus on opportunities for sustainable successes and results. The culture transformation, in effect, grows from these small change interventions that are built on the basis of new values and provide a lot of 'success' feedback.

Creating shared meaning

As indicated, to "assemble" the building blocks of an adaptive culture, you must create shared meanings, internal connectedness, and external responsiveness. To achieve this aim, leaders make sure every action is high in terms of openness, coherence, and understandability.

Openness

Openness is the quality of organizations which are effective at listening to information and using this information to enhance or improve adaptability. Open cultures can be thought of as cultures which have learned how to adapt. In this sense, 'to adapt' means to be able to make different responses to the environment in order to ensure survival and growth. Learning is an integrative experience where a change in behavior, knowledge, or understanding is incorporated into the organization's existing repertoire of behavior and schema (values, attitudes and beliefs). For example, it is possible to acquire a set of competencies that can be repeated in familiar or known circumstances; however, if adaptation has taken place, competencies can also be repeated and even adapted in unfamiliar, unanticipated situations. Thus, an open culture can be defined as one that is able to:

- Observe itself;

- Reflect on the observed differences between what is(reality) and what should be (desire);

- Understand how to change itself and its values regarding needs, practices, values, and relevant customer expectations (both internal and external);

- Successfully implement and carry out planned change.

Coherence

Coherence refers to the degree to which everything in a system hangs together and works in harmony towards common purposes and goals. In adaptive, productive cultures a high degree of coherence can be observed. Systems, structures, strategies, staff, skills, relevant science and technology, and managerial style all work together to achieve the goal of economic survival, employee well-being, and increased value for stockholders (customers, stockholders, employees). Routines and rewards are aligned with and support the overarching purposes and values inherent in the culture. Decision making, operations, and culture all work together toward shared objectives, high performance, sustained competitiveness, and economic success.

Understandability or legibility

An adaptive, productive culture is 'understandable' or 'legible'. In such a culture, behavior, performance, strategy, operations, routines and rewards, and structures are readable and understandable in terms of the vision, mission, shared values, and key goals. We have found that adaptive cultures provide:

- Focused information that is clearly and directly related to the vision and strategy of the company

- Leadership practices that are consistent with the stated purposes

- People that can relate their activities to the accomplishment of the shared goals

The total transformation process

...enterprises of great pith and moment,
With this regard, their currents turn awry
And lose the name of action.

Shakespeare, Hamlet (3.1.86)

The focal image of Part 3 is that of the crystal growing in its medium. The image is intended to reinforce the notion that cultural transformation is possible albeit rare. There are many reasons it is rare but the primary reasons are a poor strategy and a lack of leadership. With regard to strategy, it turns out the culprit is a lack of a sound people strategy, one that engages the subjective world of meaning of the people who make up the organization. With regard to leadership, it turns out the culprit is leadership practices that routinely destroy meaning, and in the process produce a loss of engagement and motivation. Here we want to explore a transformation process that is truly holistic and that engages both the minds and the hearts of the people who ultimately bring about adaptation.

Cultural adaptation and long term thinking

If culture is the DNA of the organization, questions arise as to whether the cultural DNA can be transformed and if so, what processes are required to do so. Our research suggests the answer to these questions is yes the culture can be transformed, but only if the basic building blocks of culture are changed. In the end, culture is a system of elements (both complex and adaptive) that are self-reinforcing, and to change the culture, one must change the elements and/or their relationship to one another.

These blocks contain the information that tell people what's important, what's real, and what is possible, etc. very much like the arrangement of cytosine, thymine, guanine and adenine form the basis for what we call genetic information and write the instructions that tell cells what to produce and how to produce it in support of the life system.

Unlike the genetic building blocks the cultural building blocks can be transformed by an act of will. This is because cultural DNA follows the law of consciousness which is that if you conceive it and live it, you will attain it That is to say, cultural DNA starts with our basic conceptions of who I/WE ARE and all else flow from this understanding.

Cultural adaptation requires long term thinking. It is important, then, to think about what we mean by 'long term'. How long does it take to change cultures? What are the processes we need to adopt to achieve successful cultural change? Effective cultural transformation takes a long and sustained commitment. Kotter and Heskett (1999) maintain there are three areas in which it is necessary for a leader to be able to think in the long term: creating a sustaining vision for the enterprise; creating a strategy to attain the vision, and leading transformation.

By the power of awareness we mean that the success of your transformation efforts will be determined by the degree to which you understand the multitude of meanings you have created and are reinforcing through the current culture and replacing these with meanings that create and reinforce behaviors that are aligned with the desired meanings. The question is how to bring this alignment about. Toward that end, we have created a thinking tool we call the Total Transformation Leadership Process or TTLP, which is diagrammed in Exhibit 15 (Mink, Esterhuysen, Mink & Owen, 1993). This approach has been created after many years of experience and research in implementing organizational change. The TTLP can be used to transform entire organizations involving hundreds of divisions or it can be used by a manager wishing to implement change in his or her own department.

Total refers to the need to understand transformation as a holistic process. Every organization is a system of inter-connected part or a complex adaptive human system. These parts interact in meaningful ways with one another and at any time, the parts will tend to

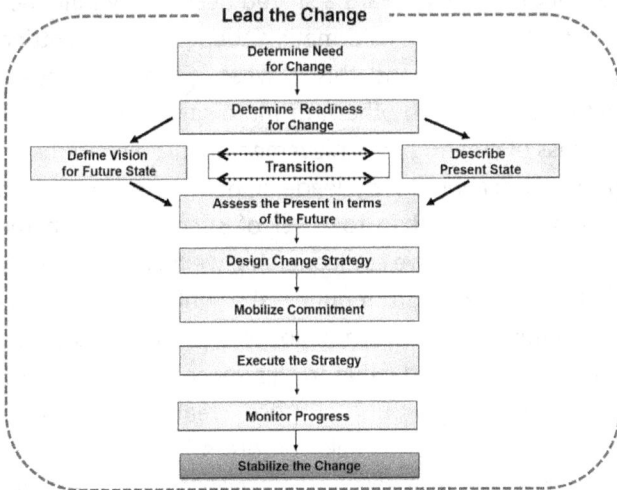

Exhibit 15 *The total transformation leadership process. © Somerset Consulting Group, 2014*

be in a state of relative equilibrium. That is, they are balanced by forces that push toward change and forces that restrain change. It is essential that change agents view change in terms of the laws governing the different systems of the organization.

Transformation refers to changing the culture. It is not about doing more of the same but about doing different things. Culture is the system output and to change that output, you have to get into the system to understand its parts and their relationship to other parts and then change the parts and/or their relationships. Transformation requires a different level or degree of awareness than is normative in most organizations.

Leadership refers to the reality that transformation is an influence process. It requires the mobilization of an array of social and structural forces and getting these to work together toward a shared destiny. By influence we mean that transformation depends on getting others mobilized and this is what leadership does— get results through others, or more specifically, it is the process of creating an achievement focused relationship based endeavor. So, by leadership we refer more to a system capacity than to a set of specific roles in the system. So when we say leadership we mean specifically that transformation requires the unleashing of the leadership potential of the social system that is in the organization. Leadership in this sense then is creating shared meaning around purpose, creating experiences that allow people to make progress in achieving that meaning (opportunities and motivation), the resources and data needed to take the right actions, and the feedback required to stay focused.

Fromm (1956) provides a useful framework for understanding what it takes to lead this process:

- Commitment;

- Discipline;

- Concentration;

- Caring;

- Understanding.

Process refers to the fact that change is not an event but a process that takes place over time. In other words, transformation requires a high level of sustained commitment over a substantial period of time. It requires learning and adjustment. Hall (1976) described the process in terms of stages of concern—the flow of feelings about transformation and he postulated three broad levels of concern:

- Personal—awareness of the need and resolution of feelings of doubt about being able to cope with a new reality.

- Task—concerns about being able to DO the change and adapt to and manage a new set of realities.

- Commitment and Innovation—concerns about improving the changes

These feelings turn out to be developmental and form a progression of stages, each of which must be resolved before a person can let go and move on.

The process

The TTLP can be thought of in terms of a series of actions that move you from understanding, new ways of thinking, new ways of acting, and finally new ways of being.

Step A: leading the transformation process— Thoughts about leadership competency

Leadership is the exercise of influence to accomplish organizational goals. In other words, leadership is getting results through others and leaders accomplish this by creating an achievement focused, relationship based environment. The ability to lead is not a given and obviously, not everyone succeeds at leadership. This is because the effectiveness of leadership behavior is determined not only by one's knowledge and skills regarding the context in which leadership is to be exercised but also by a set of underlying characteristics that influence the effectiveness with which the influence process is exercised.

Since effective leadership is a critical dimension of cultural transformation, it is imperative that we understand just what it is. First, we need to realize that the exercise of leadership reflects an intention. In some people, this intention is strong; in others it is not so strong. An intention is a desire to get work done through the exercise of influence. Second, we may or may not be aware of this intention, or rather these intentions. Some of us are motivated by a need for power, some by a need for achievement, others by a need for doing good works. Not all of these intentions are equally effective in the exercise of influence. Third, even though you may have a position of power, this does not mean that people will follow your lead. However, unless they choose to do so of their own free will, you cannot become an exemplary leader over the long haul. So, the question becomes, how does one express one's intentions so as to create followership.

The nature of leadership competency

We call these intentions to lead leadership competencies. Experts define competency as an enduring underlying characteristic of behavior that is causally related to a give outcome across time and situations (Spencer & Spencer, 1993). What this means is your competencies determine the outcomes you will achieve as a leader. By enduring we mean that these competencies tend to be typical of us across the span of our lives. By underlying, we mean that the intentions to lead are not directly observable but are rooted in our unconsciousness. By causal, we mean that your intentions create the kind and quality of followership and other outcomes you create.

There are two kinds of competency that determine how you go about the exercise of leadership: motive competencies and trait competencies. The former competencies determine the intensity of one's intentions to lead others while the latter influence how the intentions to lead are operationalized or put into play. For example, the need for achievement is a motive competency. The intensity of your need determines the intensity of your achievement related actions. By the same token, how you express this intention to achieve is also influence by your need for interaction with others.

If you tend towards being introverted this affects how you express your need for achievement; if you tend towards extraversion, this has a different effect on how you express your need for achievement.

We can put this discussion all together in terms of the graphic below (see Exhibit 16). There are three parts to this model: intents, actions and outcomes. The competency provides the psychic fuel for your actions, not only in terms of the quality of your inactions, but also in terms of their persistence in the face of difficulty, their creativity, their respect for others and so on. Neither the actions taken nor the outcomes achieved, what we normally observe, are the competency but the observable effect of the competency.

Finally, leadership is a social process. It requires the effort of others, so it is not just your competence that determines your success as a leader it is also how your efforts to influence are perceived by those whom you hope to influence. If others perceive you to be caring and trustworthy, they will follow you; if they don't they will not follow, at least without a huge cost.

The bottom line is that the outcome of a cultural transformation or of any of the steps of the process is determined by how competent a given organization's leaders are. Happily we know a great deal about what these are. We have summarized these in Exhibit 17.

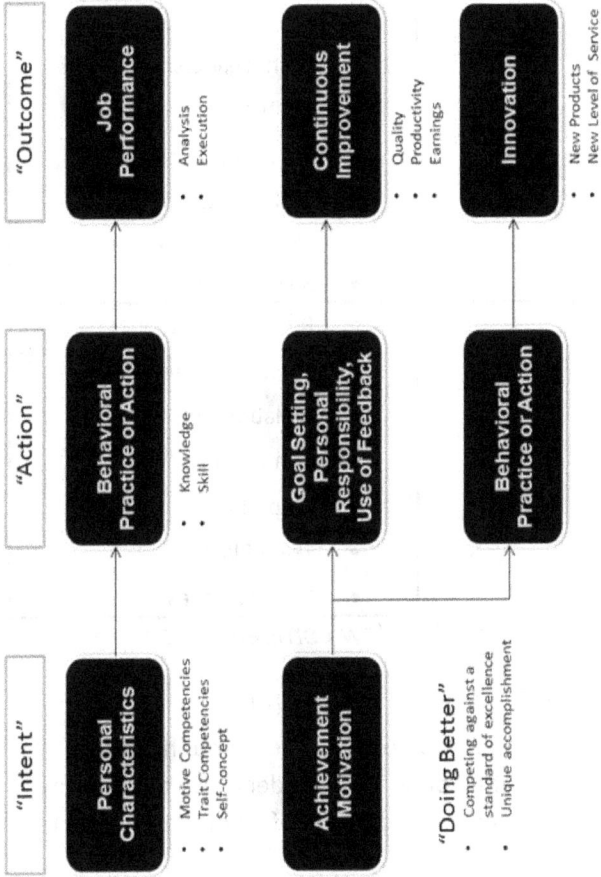

Exhibit 16 *How competencies work*

"Intent"

Personal Characteristics
- Motive Competencies
- Trait Competencies
- Self-concept

Achievement Motivation

"Doing Better"
- Competing against a standard of excellence
- Unique accomplishment

"Action"

Behavioral Practice or Action
- Knowledge
- Skill

Goal Setting, Personal Responsibility, Use of Feedback

Behavioral Practice or Action

"Outcome"

Job Performance
- Analysis
- Execution

Continuous Improvement
- Quality
- Productivity
- Earnings

Innovation
- New Products
- New Level of Service

At the _____ level	The critical competencies are:
Senior	• Openness • Decisiveness • Caring for people • Integrity • Team Leadership • Influence • Emotional Resilience • Ability to Communicate Orally • Self-Awareness
Front Line Supervisory	• Respect for others • Trustworthiness • Collaborativeness • Competence • Empathy • Team Player • Self-Awareness

Exhibit 17 *How competencies work*

The right people, the right time

Knowing the competencies is not enough; however, the issue is whether or not your leaders have the capacity to lead. The leadership capacity of an organization is built by doing two things:

- Selecting leaders that have the capacity to lead;

- Developing the specific practices that are known to be effective.

Unfortunately, many organizations do not have a process in place for selecting the right people nor for developing them. While this isn't the place to discuss these processes, it needs to be noted that both can be done in a cost effective manner. Here are the major points you will need to consider. There are six foundations that, at least to some degree, need to be in place for you to successfully select and develop leaders in your organization. These are:

- A strategic focus;

- An understanding of your competency and expertise requirements for executing your strategy;

- A process for selecting leaders with the pre-requisite competencies;

- A process of data driven feedback;

- The availability of coaching and mentoring;

- An abundance of internal support systems that encourage and reinforce learning, including an effective measurement and recognition system.

Step 1: Establish the need for change

People only change in response to a very clear need. This usually involves distress such as confusion, dissonance, and fear, or a more positive motive such as intense desire. The satiated and the comfortable are less likely to make a behavioral change no matter what others may desire. Organizational transformation usually starts with a clear cut need or threat to survival. Needs are the gaps between current and required performance which are not equivalent.

Your role as an agent of transformation is to mobilize and harness the energy inhering in this need. Keep in mind that the current need is an outcome of the current culture and that you cannot get a different result by trying harder; rather you must create a different set of expectations and meanings (mind-sets) and provide the experiences that allow people to reinforce these new meanings. Unless you communicate this clearly, forcefully, behaviorally, and consistently, you cannot succeed in communicating the need for transformation.

Step 2: Define the future state—Establishing vision, mission, and values

The most fundamental step in transforming a system is the development of a shared vision. The vision provides a statement of the fundamental purposes of the organization and the values by which it is to achieve the vision. The vision forms the new meanings toward which transformation is pointed; it provides the tension and energy needed to sustain the change effort over time, and provides the framework for evaluating the success of efforts to effect change. It serves to align people, motivate them, and provide them a sense of control over their own destinies.

More importantly, the vision communicates to people the meaning of the transformation process and provides the framework for aligning meaning, experiences, actions, and results. In the end, every behavior must communicate a consistent meaning. In short it must exhibit:

- Legibility;

- Coherence;

- Understandability;

- Consistency.

These meanings define the foundational relationships in the ecospace the organization inhabits. If leaders want their employees and customers to be more loyal to the organization, new meanings required to achieve this (since the current ones are not working). What new mind-sets and values must the leaders of the organization define, communicate, live, and hold people accountable for? How will the organizations leaders demonstrate that they care for employees and customers, and that the leadership are competent to carry out the vision? Can the leaders walk the talk clearly and consistently? What new behaviors must be exhibited? What old behaviors must be extinguished?

The goal of transformation is to keep the four building blocks aligned and in harmony with the external environment. We have found that answering four questions enables us to define the future state in meaningful terms:

1. What specific results do we want to achieve?

2. What mind-sets and beliefs support the accomplishment of these results and enable people to become fully enrolled in the process of adaptation and change?

3.   What specific experiences must we create to support these beliefs and provide people the opportunity to contribute to success?

4.   What specific actions must people be able and equipped to take in order to succeed?

Step 3: Describing the present

Keep in mind that we have defined the organization as a set of spaces in which people are more or less empowered to meet their needs.   In other words, these spaces define a set of relationships the individual experiences each and every day. These experiences in turn determine the quality of that person's inner work life. When they are positive, so is a person's inner work life; when they are negative, so is a person's inner work life. The reason you have a need for transformation is that in some significant way, the quality of inner work life is negative.

Fundamental to the change process is an understanding of where the organization is now.   How must the current spaces be changed and rearranged so as to create the opportunity for daily engagement in vision relevant experiences?   Space is where your people encounter opportunities or barriers for making progress in creating meaning. In our studies we have used the CultureGRADE® (see Attachment 1), though there are other processes available for determining this. Again we use the following four questions to assess the current state of the system:

1. What specific results are we now achieving?

2. What are the specific mind-sets and beliefs that have produced the current results?

3. What specific experiences are reinforcing these mind-sets and beliefs and producing observed results?

4. What specific actions are people engaged in that must be changed? How must employees be supported and equipped to take in order to succeed?

Step 4: Assess the present in terms of the future

The critical question here is how you improve employees' opportunities to make progress at creating meaning each and every day. The work space, symbolic, virtual and physical, is where employees encounter these opportunities. So, change leaders must make sure they know how to create spaces that provide employees multiple and daily opportunities to make progress in creating meaning. This means that the nature of the space depends on the fit for purpose/result role occupants are expected to produce.

This task is critical for it answers the question what do we need to do more or and less of to bridge the gap between where we are and where we need to be? Exhibit 18 illustrates the basic challenge. The current space is creating the current results. To create new results, what must change in the present space?

Assess the present in terms of the new vision

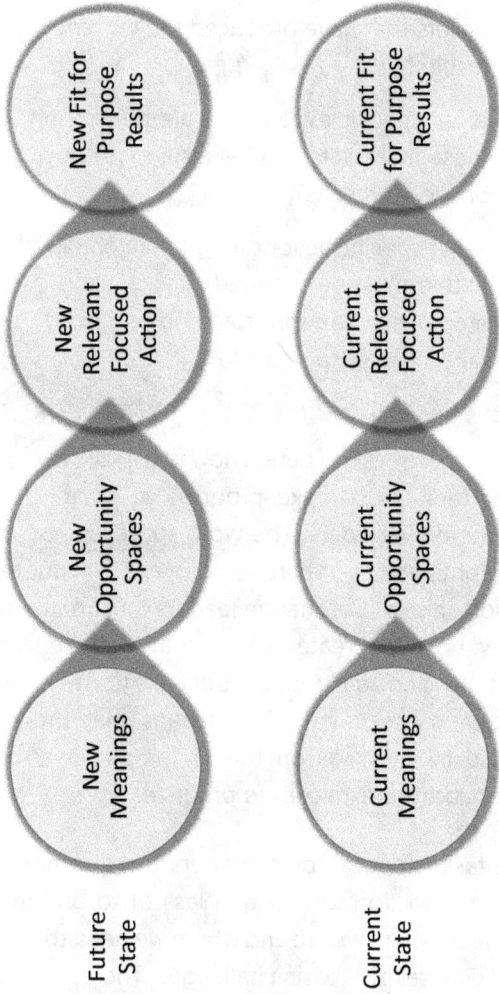

Future State

New Meanings → New Opportunity Spaces → New Relevant Focused Action → New Fit for Purpose Results

Current State

Current Meanings → Current Opportunity Spaces → Current Relevant Focused Action → Current Fit for Purpose Results

Exhibit 18 *Change is a choice*

Step 5: Act at three levels

People who work for an organization generally do what they believe they are expected to do. An expectation is a mind-set to get new levels of energy and commitment from employees; leaders must create the spaces in which those expectations can be fulfilled (Olson & Eoyang, 2001). These spaces are the place where people get to engage in new ways of being and doing, new ways that enable them to take actions that result in progress toward creating meaning and meeting their needs. You might say that these spaces should ramp up their level of focused activity.

Taking concrete and specific action is the next step in the transformation process. There can be no change in BEING without a change in DOING. Everyone knows how to plan. What we can offer here is that the plans must take into account several factors. Plans must:

- Allow for individual contribution to the success of the change effort;

- Be aligned so that every department is working towards the same goal;

- Assist the implementation teams to experience short term success.

Intervening at the individual, group, and organizational levels is the heart and soul of the change process. One of the key ingredients for the success of a change is to identify and target some immediate wins. These short term successes build motivation and commitment to the change process. At the group level, we focus on building strong, fully functioning work teams by optimizing the group process by developing clear purposes, developing quality relationships built on trust and honesty, problem solving, and team learning. At the organizational level, we focus on integrating and aligning business systems and on creating performance measurement systems that incorporate the organization's vision, mission, values and philosophies, and key goals.

Step 6: Make the change happen

Transformation is generally effected through the use of various levels of continuous improvement teams, each working in a coordinated way to bring about change. Four kinds of teams are generally involved in implementing and managing the transition:

- Leadership transformation teams which make sure all the steps that are taken are aligned around the shared meanings, experiences, actions, and results that you want to create.

- Continuous or process improvement teams which tackle process improvement goals.

- Infrastructure teams which tackle issues related to developing systems which support the transformation.

- Transition management team which serves as an internal coordinating team to keep all change projects on track and working in harmony.

Conditions for optimal success in the change are identified, the levels of use of the change are analyzed, and interventions related to use level are suggested.

Step 7: Stabilize the transformation

Plans most often fail because leaders do not understand the crucial role of measuring the plan's performance relative to expectations. To keep an organization from reverting to its former state, it is important for leadership to stabilize or refreeze the organization at a new level of performance. Organizations must be sensitive to any constraints that stand in the way of maintaining a change, and must be ready to remove them on an

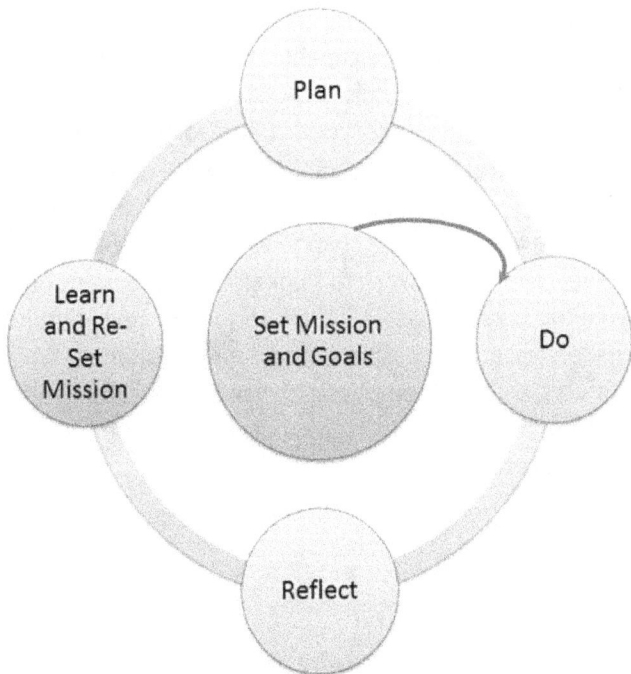

Exhibit 19 *The plan, do, reflect, learn cycle*

ongoing basis. The PDRL cycle illustrated in Exhibit 19 suggests a model of thinking about change and for stabilizing its effects.

Conclusions

The reality of dealing with complex social systems is that creating an adaptive culture is critical to their success. The flip side of this reality is that the goodness of fit of any organization's culture will be in a constant state of change, leading to the emergence of organized forces which mitigate against success but increase the likelihood of failure. As a leader, then, you must be consistently mindful of the degree to which meanings, experiences, actions and results are aligned and supportive of your vision and you must be capable of taking action quickly in order to maintain this level of alignment. In short, you must actively lead your culture or it will end up leading you. Clearly, the highly successful organizations of today and the future will be capable of this kind of adaptive learning. This change will be accomplished by leaders which learn how to think from a full dimensional systems perspective. Cultural alignment and mindful leadership are two aspects of adaptive cultures. Such cultures, with their emphasis on values as well as quality will be the market and community leaders in the future global economy.

Iron has memory, so do organizations.

Rocks can breathe, so do organizations.

Crystals can learn; so can organizations.

References

3m Company (2002). A Century of Innovation: The 3M Story. 3m Co.

Ackoff, R.L., (1981) Creating Your Corporate Future. New York: John Wiley & Sons

Amabilie, Teresa and Kramer, Steven (2011). The Progress Principle: Using Small Wins to Ignite Joy, Engagement, and Creativity at Work. Kindle E-Book.

Amabile, T. and Kramer, S. (2012). How Leaders Kill Meaning at Work. McKinsey Quarterly Online, January, 2012.

Argyris, C., & Schön, D. (1978). *Organizational learning: A theory of action perspective.* Reading, MA: Addison-Wesley.

Aronson, E. (2008). The theory of cognitive dissonance: A current perspective. In L. Berkowitz (ed.) Advances in experimental social psychology. New York, New York: Academic Press, Vol 4, pp. 1-34.

Ashby, W. R. (1968). Variety, Constraint, and the Law of Requisite Variety. In Buckley, W., Ed., *Modern System Research for the Behavioral Scientist: A Sourcebook.* Aldine Publishing Company: Chicago.

Bandura, Albert (1997). Self-Efficacy: The Exercise of Control. Worth Publishers.

Barthes, R. (1976) Mythologies. U.K.: Paladin.

Bennis,W. (1984) The 4 Competencies of Leadership. Training and Development Journal.

Chatwin, B. (1987) The Songlines. U.K.: Picador.

Calvino, Italo, Invisible Cities. Harcourt, Brace, and Jovanovich, 1978.

Deal, T. and Kennedy, J. (1982) Corporate Culture: The Rites and Rituals of Corporate Life. Reading, Mass: Addison Wesley.

Dietz, A. & Mink, O. (2005). Police systems and systems theory: An interpretive approach to understanding complexity. *Journal of Police and Criminal Psychology, 20,* 1, 1-15.

Fromm, E. (1956). The Art of Loving.

Gardner, M. & Ashby, R. (1970). Connectance of large dynamic (Cybernetic) systems: Critical values for stability. Nature, 228, 784.

Hall, E. (1976) Beyond Culture. Garden City, New York: Anchor Press.

Hampden-Turner, C. (1990) Creating Corporate Culture. Reading, MA.: Addison-Wesley Publishing Co.

Hess, Edward, D. (2010). *Smart Growth*, Columbia Business School Publishing, New York

Kotter, J.P. and Heskett, J.L. (1992) Corporate Culture and Performance. New York: The Free Press.

Land, G. and Jarman, B. (1992) Break Point: Mastering the Future Today. New York: Harper Business.

Mink, O., Esterhuysen, P., Mink, B. & Owen, K. (1993). Change at Work: A comprehensive management process for transforming organizations. Jossey-Bass: NJ

Mink, O. Mink, B. and Owen, K. (1987) Groups at Work. New Jersey, Educational Technology Press.

Mink. O. & Owen, K. (1992) Analyzing Team Effectiveness Technical Manual.

Mink, O., Owen, K. and Bright, S. (1993). Iron has memory, Rocks breathe slowly, Crystals learn: Long Term Thinking and Cultural Change. Asia Pacific Journal of Quality Management, 2, 2, 26-39.

Olson, E. & Eoyang, G. (2001). Facilitating Organization Change: Lessons from Complexity Science. SanFrancisco, CA. Pfeiffer.

Owen, K. O. & Dietz, A. S. (2012). Understanding organizational reality: Concepts for the change leader. *SAGE Open,* 2012, 2, 1-14: DOI: 10.1177/2158244012461922.

Owen, K., Northcutt, N. & Dietz, A. (2013). Engaging employees in interpreting survey results: Using PathMAP® as a tool to drive understanding and change. *SAGE Open* January-March 2013, 3, 1-16: DOI: 10.1177/2158244013481478.

Peters, T. & Waterman, B. (1982) In Search of Excellence. New York: Harper & Row.

Pink, D. (2011). Drive: The Surprising Truth about What Motivates Us. Kindle Books.

Read, B., (2010). Personal Interview. Safety Leaders Group.

Reber, P. (2010). What is the memory capacity of the human brain? *Scientific American,* April 19, 2010.

Salem, P. (2008). The seven reasons organizations do not change. *Corporate Communications: An International Journal, 13,* 333-348.

Schein, E. (1990). Organizational Culture: What It Is and How To Change It. In P. Evans, et al., Eds. Human Resource Management in International Firms. New York, Saint Martin's Press.

Schein, E. H. (2010). Organizational Culture and Leadership. John Wiley and Sons. Kindle Edition

Senge, P. (1990). The Fifth Discipline: The Art and Practice of the Learning Organization. New York: Doubleday.

Sherman, D. & Cohen, G. (2006). *The psychology of self-defense: Self-affirmation theory. In Mark Zanna (ed.)* Advances in Experimental Social Psychology, New York: Academic Press.

Shakespeare, Hamlet (3.1.86).

Spencer, L.M. and Spencer, S. M. (1993) Competence at Work: Models fore Superior Performance. New York, NY: John Wiley and Sons.

Tavris, C. and Aronson, E. (2007). Mistakes were made (But not by me): Why we justify foolish beliefs, bad decisions, and hurtful acts. Orlando, FL: Harcourt.

Trice, H.M. and Beyer, J.M. (1993) The Cultures of Work Organizations. Englewood Cliffs, N.J.: Prentice Hall.

Warhol, A. (1977). The philosophy of Andy Warhol: From A to B and Back Again. Orlando, FL: Mariner Books.

Winter, J., Owen, K., Read, B., Ritchie, R. (2010). How Effective Leadership Practices Deliver Safety Performance AND Operational Excellence: A Case Study. SPE Oil and Gas India Conference and Exhibition, Mumbai, India, 20–22 January 2010.

Attachment 1

CultureGRADE® provides an easy to use yet valid way of looking at this question as it measures the quality of the important relationships in your organization, namely:

- Relationship with senior management
- Relationship with one's supervisor or one-up manager
- Relationship with peers
- Relationship to job role
- Relationship to risk and safety
- Relationship to performance management systems

CultureGRADE® provides a grade for each of these critical relationships as well as a framework for figuring out what you must start, stop, and/or continue doing.

Again we use the following four questions to assess the current state of the system?

1. What specific results are we now achieving?

2. What are the specific mind-sets and beliefs that have produced the current results?

3. What specific experiences are reinforcing these mind-sets and beliefs and producing observed results?

4. What specific actions are people engaged in that must be changed? Must people be able and equipped to take in order to succeed?

GRADE	A	B	C	D	F
TYPE	Adaptive	Paternal	COMPLIANT	Manipulative	Exploitative
Shared MEANING	Personal enrollment and self-motivation	Agreement with rules, reward and motivation	Tolerance of rules inconsistent reward	Lip service only	None
INTERNAL Connectedness	Self-defining efficacy and optimism	Rule-defining and acquiescence	Tolerance	Apathy	Helplessness
EXTERNAL Responsiveness	"Self-driven" Motivated effort	"Pushed" Motivated effort	Moderate motivated effort	Low Motivated effort	No Motivated effort

Type	What Drives Behavior	Responsibility	Accountability
Adaptive	Values Intrinsic Promotes collective efficacy	Mutual Interdependent Acknowledge Create Empower	The team Shared fate Committed to a shared purpose
Paternal	Rules Extrinsic Promotes individual efficacy	Systems Competence	The individual
Compliant	Control Oder Prescription Extrinsic Promotes conformity	Systems and processes	The Supervisor
Manipulative	Reactive Extrinsic Promotes avoidance		Diffuse
Exploitative	Survival of the fittest Extrinsic Promotes fear	Supervisor	No one Everyone out for self

About the authors

Keith Owen, Ph.D. cofounded Somerset Consulting Group has been doing organizational research and consulting for 30 years. He has been involved in the design, analysis, and implementation of 100's of employee and customer research studies in a variety of organizations, including oil companies, high tech companies, utilities, hospitals, and manufacturing organizations in both the public and private sectors. He received his doctoral degree in Experimental and Personality Psychology from the University of Texas. His strengths include research design and implementation, data analysis and data modeling to establish cause and effect relationships, change management and organizational development often using research as a catalyst for driving change.

keith@somersetcg.com

A. Steven Dietz, Ph.D. is currently an Assistant Professor in the Department of Occupational, Workforce, and Leadership Studies at Texas State University. He is an experienced organization dynamics consultant working with both private and public organizations over the past 20 years. He has extensive experience with police departments and military organizations, and was the Director of the Texas Institute for Public Problem Solving (TIPPS) from 1997-1999. Steven is also a veteran of the U. S. Navy and currently serves in the Texas Army National Guard.

scarver103@gmail.com

Robert 'Skip' Culbertson is currently the senior Director of Custom Executive Education Programs and Thought Advancement at the Darden School of Business at the University of Virginia in Charlottesville, Virginia. Previously, Skip enjoyed 30 years in leadership positions in the energy sector and as a management consultant and executive coach. He has served as Managing Director of the Blue Sky Consulting Group. Skip enjoys travel, art, and searching for antique treasures in his spare time.

skipculbertson@gmail.com